5th International Conference on Advances in Nuclear Science and Engineering (ICANSE-2015)

Journal of Physics: Conference Series Volume 799

Bandung, Indonesia
18 – 20 November 2015

ISBN: 978-1-5108-3718-8
ISSN: 1742-6588

Printed from e-media with permission by:

Curran Associates, Inc.
57 Morehouse Lane
Red Hook, NY 12571

Some format issues inherent in the e-media version may also appear in this print version.

Printed by Curran Associates, Inc. (2017)

For permission requests, please contact the Institute of Physics at the address below.

Institute of Physics
Dirac House, Temple Back
Bristol BS1 6BE UK

Phone: 44 1 17 929 7481
Fax: 44 1 17 920 0979

techtracking@iop.org

Additional copies of this publication are available from:

Curran Associates, Inc.
57 Morehouse Lane
Red Hook, NY 12571 USA
Phone: 845-758-0400
Fax: 845-758-2633
Email: curran@proceedings.com
Web: www.proceedings.com

Table of contents

Volume 799

International Conference on Advances in Nuclear Science and Engineering 2015
18–20 November 2015, Bandung, Indonesia

Accepted papers received: 21 December 2016
Published online: 28 February 2017

Preface

Papers

PREFACE

The Fifth International Conference on Advances in Nuclear Science and Engineering (ICANSE-2015) has been successfully conducted in Bandung Institute of Technology campus, Bandung, Indonesia, at 18-20 November 2015. This conference has been organized by Bandung Institute of Technology (ITB, Indonesia). This conference was also supported by Indonesian Journal of Physics (IJP). This conference aims at summarizing recent research activities relevant to the advanced development of application in the nuclear science and engineering and facilitate communication among relevant experts.

About 100 persons from Indonesia, Japan, Turkey, Korea, and Rusia have been participated in this conference. About 70 presentations including 8 invited talk were presented. The presentations are grouped into 7 areas of particular interest: (1) : Innovative NPP, (2) Fuel Cycle and Waste Management, (3) Simulation and Computation, (4) Radiation and Biophysics, (5) Thermal Hydraulic and safety, (6) Nuclear Physics, and (7) Nuclear Materials and general subject related to nuclear.

From about 40 full papers submitted after the conference for publication in IOP conference proceedings, after peer review process by Relevant Experts, 29 papers were accepted for publication in this proceedings. We are indebted to all of authors for submiting their original papers.

We would like to thank all members of International Board, members of Organizing Committee, and my gratitude to all those who help the success of this conference.

Zaki Su'ud

ICANSE 2015 Chairman

Dwi Irwanto

ICANSE 2015 Co-Chairman

Peer review statement

All papers published in this volume of *Journal of Physics: Conference Series* have been peer reviewed through processes administered by the proceedings Editors. Reviews were conducted by expert referees to the professional and scientific standards expected of a proceedings journal published by IOP Publishing.

A Roadmap of Innovative Nuclear Energy System

Hiroshi Sekimoto

Tokyo Institute of Technology

Email: hsekimot@gmail.com

Abstract. Nuclear is a dense energy without CO_2 emission. It can be used for more than 100,000 years using fast breeder reactors with uranium from the sea. However, it raises difficult problems associated with severe accidents, spent fuel waste and nuclear threats, which should be solved with acceptable costs. Some innovative reactors have attracted interest, and many designs have been proposed for small reactors. These reactors are considered much safer than conventional large reactors and have fewer technical obstructions. Breed-and-burn reactors have high potential to solve all inherent problems for peaceful use of nuclear energy. However, they have some technical problems with materials. A roadmap for innovative reactors is presented herein.

1. Introduction

The Society of Chemical Engineers, Japan recently published Energy Technology Roadmaps of Japan - Future Energy Systems Based on Feasible Technologies Beyond 2030. I was in charge of writing the part of Nuclear Power Generation [1]. Though this roadmap is a Japanese roadmap, the main part of it discusses the relations between nuclear developed countries and developing countries, and the potentials of innovative nuclear reactors solving the problems we are facing. The present paper is written by modifying the above Roadmap of Japan and tried to make general discussions on the worldwide researches and developments of innovative nuclear reactors appropriate to the International Conference on Advances in Nuclear Science and Engineering.

Global warming is an increasingly urgent problem in the world. Nuclear power is one of the few technologies that can mitigate climate change by reducing CO_2 emission [2]. After the slumped period and the 21st century started, the number of nuclear reactors in the world started to increase. However, the Fukushima-Daiichi accident greatly influenced the atomic energy programs for many countries. Now the number of nuclear reactors starts to increase again, but the public acceptance becomes much more difficult than before in many countries.

Nuclear energy not only has a safety problem but also several difficult problems caused by generated radioactive materials and technologies in common with nuclear weapons production. However, it has unique merits such as extremely high energy density, abundant resources, and zero CO_2 emissions.

According to a nuclear energy roadmap from international agencies [3], the capacity of nuclear energy will not change from the present through 2050 in OECD countries, but will grow rapidly in developing countries. Developed countries should help developing ones successfully use nuclear energy. Safe, simple and easy reactors are required for developing countries.

Some small reactors (SRs) have these features, although their economy is a difficult challenge to overcome. Breed-and-burn reactors (B&BRs) have high potential to overcome all the above problems, but several technical problems await solution.

The following section describes fundamental characteristics of nuclear energy. A brief history and present status are presented in Section 3, and prediction of nuclear energy by international organizations is addressed in Section 4. Innovative nuclear energy is the topic in Section 5, where we address SRs and B&BRs. A technology roadmap is presented in Section 6, and benefits and future vision are addressed in Section 7.

2. Fundamental Characteristics of Nuclear Energy

Humans (not modern humans but Homo erectus) started to use/control the fire on woods about a million years ago. The wood fire is considered as a stored energy, and we can obtain and use whenever we need. It was a turning point in the cultural aspect of human evolution that allowed humans to cook food and obtain warmth and protection. It developed civilizations. The scale of this use and its environmental effects were small. Humans began to use another stored energy, fossil fuels, on a large scale several hundred years ago and this made possible the industrial revolution and modern civilization. Huge amounts of this energy can be used whenever required. Their consumption rate increased rapidly, and resource and environmental problems have become urgent issues.

Nuclear energy, which is also a stored energy, was discovered in the middle of the 20th century. The important feature of the stored energy is its density (specific energy). A list of energy density for several fuel materials is shown in table 1 [4], where one can see a large difference (about a million times) of energy density between nuclear and fossil fuels. Compared with fossil fuels, nuclear energy has several merits in addition to the greater density, such as no CO_2 emission and no oxygen requirement.

Table 1. Energy density [4]

Fuel material	Energy type	Specific energy (MJ/kg)	Direct uses
Uranium (in breeder)	Nuclear fission	80,620,000	Electric power plants (nuclear reactors)
Thorium (in breeder)	Nuclear fission	79,420,000	Electric power plants (nuclear reactors)
Hydrogen (compressed at 70 MPa)	Chemical	142	Rocket engines, experimental automotive engines
LPG (including Propane / Butane)	Chemical	46	Cooking, home heating, automotive engines
Jet fuel	Chemical	43	Aircraft
Fat (animal/vegetable)	Chemical	37	Human/animal nutrition
Coal	Chemical	24	Electric power plants, home heating
Carbohydrates (including sugars)	Chemical	17	Human/animal nutrition
Protein	Chemical	17	Human/animal nutrition
Wood	Chemical	16	Heating, outdoor cooking
TNT	Chemical	4.6	Explosives
Lithium battery (non-rechargeable)	Electrochemical	1.8	Portable electronic devices, flashlights
Alkaline battery	Electrochemical	0.67	Portable electronic devices, flashlights
Lead-acid battery	Electrochemical	0.17	Automotive engine ignition

The means for releasing nuclear energy were discovered during the World War II. One of the main objectives of this war was to secure energy, especially oil. Nuclear energy became another energy resource. Figure 1 shows the resource amount and available period for each energy resource, estimated by the author more than 20 years ago [5]. Resource amounts for this graph were obtained from data available at that time. The values may be considerably different today, but the differences are sufficiently small for the present discussion. The available years were obtained by dividing resource amounts by total annual energy consumption in the future, 0.32 ZJ/year (1 ZJ = 10^{21} J). Because several energy types are typically used in a mixed manner, actual available periods are considered to be longer than the values shown.

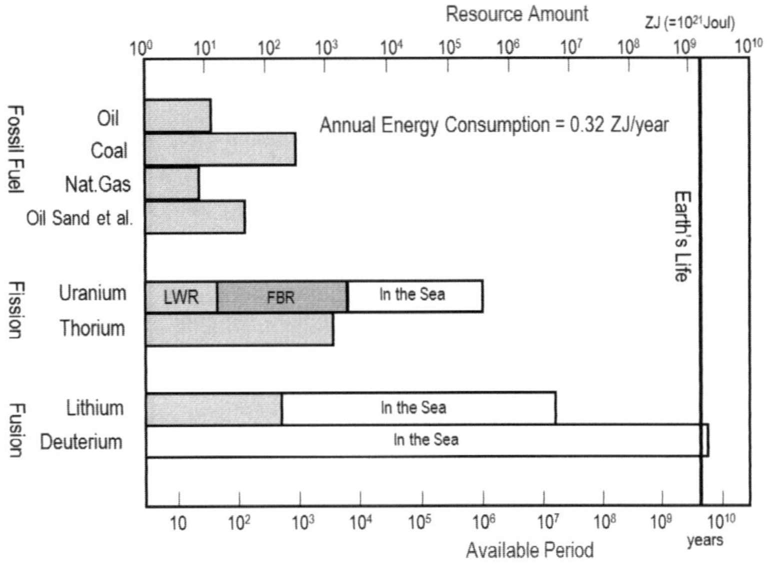

Figure 1. Estimated resource amount and available period for individual energy resources [5]

There are dwindling supplies of fossil fuels. Even for the nuclear energy, if only light water reactors (LWRs) with uranium fuel are used, nuclear energy sources can also provide only short-term supply, since LWRs can utilize only less than 1% of natural uranium.

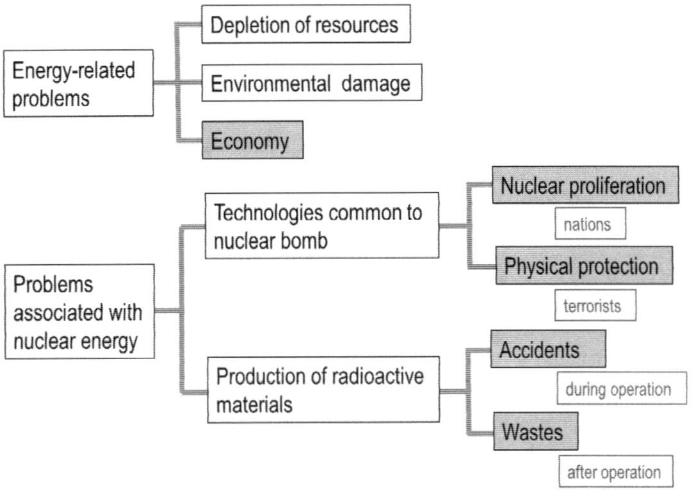

Figure 2. Necessary and sufficient requirements for nuclear energy systems [7]

Nuclear reactors produce substantial radioactive materials during operation. This can cause devastating accidents during reactor operation and a problem with radioactive wastes, even after operation cessation. Another inherent problem of nuclear energy is that it uses materials and technologies that are closely related to atomic bomb production. Problems associated with these bombs require implementation of measures for safeguards, terrorist threats, and nuclear proliferation. In this work, we simply call these problems nuclear threats [6]. Cost effectiveness is always an important requirement for energy. Thus, for nuclear energy to be used as a primary energy, all problems associated with a) limited resources, b) safety, c) waste disposal, d) nuclear threats, and e) cost must be solved (figure 2) [7].

For a), the resource problem, fast breeder reactors (FBRs) can use almost all natural uranium. It can extend the fuel available period by almost a hundred times. It also increase the value of natural uranium by almost a hundred times. It means expensive uranium such as uranium in the sea water becomes available as a cost competitive fuel. Then the available period can be extended by almost 10,000 times (figure 1). It is considered enough for humans. For b), the safety problem, after the Three Mile Island (TMI) accident, several SR concepts were proposed [8], which are addressed in detail in Section 5.1. For c), the waste problem [9], at present only underground disposal is promoted in most countries having operational nuclear reactors. However, it is difficult to obtain public acceptance in many countries, and innovative technologies such as transmutation of long-life radioactive wastes are under study. Another creative solution to this problem is proposed in Section 7. The nuclear threats problem, d) [6], is related to peaceful use of nuclear energy, but the relationship is not simple. Even if peaceful use ends, the risk of nuclear threats will persist. If care of nuclear materials and technologies is inadequate, the situation will become worse than the present peaceful use situation. We should develop a peaceful use system that strongly protects against this problem. Regarding e), the cost problem is common to most issues related to daily life, and this is strongly tackled even at present.

3. Brief History and Present Status

Nuclear energy was discovered during study on basic physics in the 20th century. In the beginning, release and utilization of this energy by humans were considered impossible. However, fission was discovered in 1939, the possibility of sustainable chain reaction of neutron-induced fission was confirmed, and the criticality of CP1 was achieved in 1942. Since then, many nuclear reactors have been constructed and operated, for military purposes in the beginning and later for peaceful uses [10]. The development of nuclear power began with 6 MWe graphite-moderated water-cooled reactor ONPP which was constructed in 1951 and started to produce electricity to supply to a conventional transmission grid in1954 in the USSR, followed by 60 MWe Shippingport PWR which was constructed in 1954 and began generating electricity in 1957 in the USA.

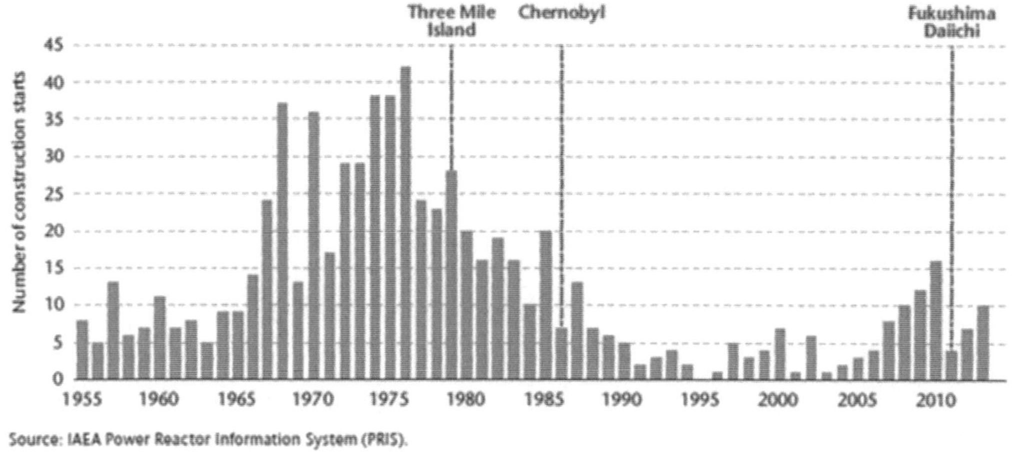

Figure 3. Nuclear reactor construction starts, 1955–2014 [3]

The number of construction starts for nuclear reactors every year from 1955 until 2014 is shown in figure 3 [3]. The number generally increased from the beginning through around 1975, when the contribution of the USA was dominant, then decreased to nearly zero around 1995. Since the mid-2000s, the number of reactors constructed has been trending upward, largely because of rapid development in China. Years of the three major accidents (TMI, Chernobyl, and Fukushima) are also shown in figure 3. Despite the serious damage from the Fukushima accident, the recent upward trend appears sufficiently strong for increase in the future.
Most reactors recently constructed are PWRs and their power level are more than 1000 MWe.

4. Nuclear Energy Prediction by International Organizations

The Nuclear Energy Agency (NEA) and International Energy Agency (IEA) recently published "Technology Roadmap - Nuclear Energy" [3] in which they show electricity production by technology in the 6 °C and 2 °C scenarios (6DS & 2DS) (figure 4). Although the renewable and nuclear contributions are not substantial in the 6DS, they become so in the 2DS. The large share of renewables in 2DS significantly changes the nuclear operating environment, since the power output of renewables changes drastically even in one day. Nuclear power is traditionally operated to meet baseload demand. It is difficult for both PWR and BWR to execute rapid load-following to meet the changes brought by renewables. New designs are required for this purpose.

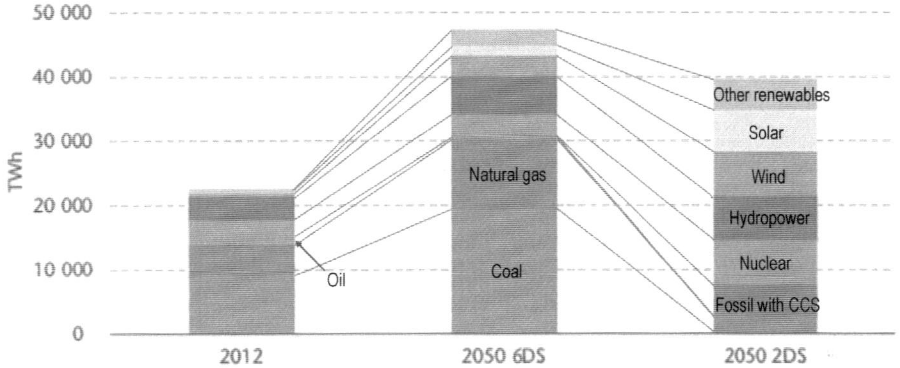

Figure 4. Electricity production by technology in 6DS and 2DS [3]

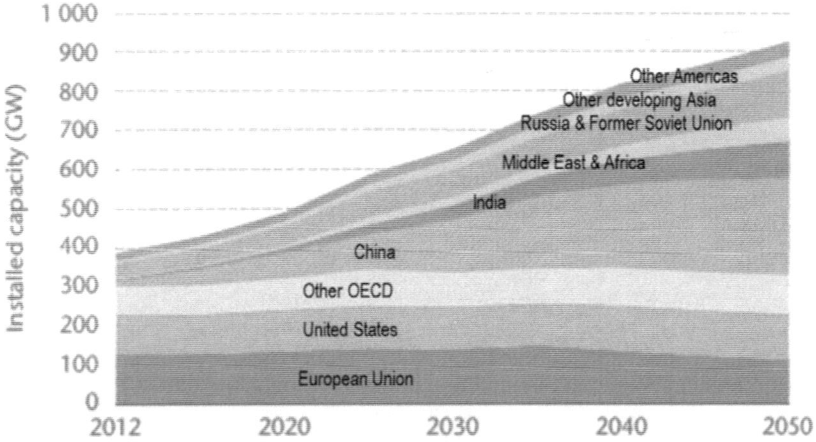

Figure 5. Nuclear generation capacity in the 2DS by region [3]

The NEA and IEA roadmap also shows nuclear generation capacity variation from 2012 to 2050 in the 2DS by region (figure 5), during which the capacity of emerging countries, especially China, changes drastically. However, the capacity of OECD countries barely changes. Current reactors have been elaborated in developed countries for use in those countries. These countries should support developing countries in the use of nuclear energy. However, much safer, simpler and easier reactors are required for developing countries.

5. Innovative Nuclear Energy

5.1. Small Reactors (SRs)

Soon after the TMI accident, many innovative reactor designs were proposed with inherent safety features, which depend not on the intervention of humans or electromechanical devices but instead on immutable and well-understood laws of physics and chemistry [11, 12]. These features are usually attained from favorable reactivity feedback for passive reactor shutdown, passive core cooling against decay heat, and superior confinement performance of radioactive materials. In general, SRs can perform these safety functions relatively easily [13].

Table 2. Small reactors (SRs) [8, 14–16]

Name	Capacity	Type	Developer
CAREM	27-100 MWe	PWR	CNEA + INVAP, Argentina
MRX	30-100 MWe	PWR	JAERI, Japan
KLT-40S	35 MWe	PWR	OKBM, Russia
NuScale	45 MWe	PWR	NuScale Power + Fluor, USA
Flexblue	50-250 MWe	PWR	Areva TA, France
SMART	100 MWe	PWR	KAERI, South Korea
ACP100	100 MWe	PWR	CNNC + Guodian, China
NP-300	100-300 MWe	PWR	Areva TA, France
IRIS	100-335 MWe	PWR	Westinghouse-led, international
CAP-150	150 MWe	PWR	SNERDI, China
mPower	150-180 MWe	PWR	B&W + Bechtel, USA
SMR-160	160 MWe	PWR	Holtec, USA
Westinghouse SMR	225 MWe	PWR	Westinghouse, USA
VK-300	300 MWe	BWR	Atomenergoproekt, Russia
PHWR-220	220 MWe	HWR	BARK, India
HTTR	30 MWt	HTR	JAEA, Japan
PBMR	165 MWe	HTR	Escom, South Africa, et al.
HTR-PM	2x100 MWe	HTR	INET + HSNPC, China
SC-HTGR (Antares)	250 MWe	HTR	Areva, France
GT-MHR	285 MWe	HTR	GA + Minatom, USA-Russia
4S	10-50 MWe	FNR	Toshiba, Japan
SVBR	10-100 MWe	FNR	AKME (Rosatom), Russia
Hyperion Power Module	25 MWe	FNR	Hyperion Pwr Gen, USA
ENHS	50 MWe	FNR	UCB, USA
LSPR, PBWFR	50-150 MWe	FNR	TokyoTech, Japan
ALFRED	120-600 MWe	FNR	Ansaldo, Italy
EM^2	240 MWe	FNR	GA, USA
BREST	300 MWe	FNR	RDIPE, Russia
S-PRISM	311 MWe	FNR	GE-Hitachi, USA
FUJI, miniFUJI	10, 100-200 MWe	MSR	IThEMS, Japan-Russia-USA
IMSR	45 MWe	MSR	Terrestrial Energy, USA
Mk1 PB-FHR	100 MWe	FGR*	MIT + UCB + UWM, USA
Leadir-PS100	36 MWe	LGR**	Northern Nuclear, Canada

*flibe cooled graphite moderated reactor, **lead cooled graphite moderated reactor

Now more than a half century has passed since Shippingport reactor started its operation, and the time is approaching of the replacement of many nuclear power reactors especially in USA. This situation was often called Nuclear Renaissance by nuclear promoters. The next generation of reactors is called Generation IV [17]. These reactors should satisfy very severe safety requirements. Especially after the Fukushima accident (Reactor type is BWR.), this requirement is inevitable. In these situations SRs are attracting strong interest. Many SRs are proposed now as shown in table 2 [8, 14-16]. Though many SR concepts are proposed, they are still in a R&D stage now. They hold a lot of merits, but they have a difficult problem of economy. The economy of these reactors is considered generally worse than the conventional large reactors because of their scale demerit. The mass-production in a factory may solve this problem.

Table 2 lists many types of SRs, especially for PWR, high-temperature reactor (HTR) and fast neutron reactor (FNR). Categorizing these by their generation, PWR, BWR and heavy water reactor (HWR) belong to Generation III and the others to Generation IV [17]. The Generation IV International Forum (GIF) was created in 2000. The GIF selected six systems as Generation IV technologies: gas-cooled fast reactor (GFR), lead-cooled fast reactor (LFR), molten salt reactor (MSR), sodium-cooled fast reactor (SFR), supercritical water-cooled reactor (SCWR), and very high-temperature reactor (VHTR). Using the type in table 2, VHTR belongs to HTR, and GFR, LFR and SFR to FNR. According to the GIF 2014 Technology Roadmap Update for Generation IV Nuclear Energy Systems [18], VHTR moved from the viability to performance phase in 2010, and SFR and LFR should do so by the end of 2015.

Technical feasibility of the SRs of PWR, BWR, HWR, HTR and some FNR (SFR and LFR) can be expected to be proven earlier. However, because of their scale disadvantages, the economy of these reactors is considered to be worse than conventional large reactors [18]. Means to overcome this obstruction are addressed in Section 6. The U.S. Energy Department is supporting R&D on SRs [19]. The IAEA is supporting long-term international activities regarding SRs.

The International Conference on Advances in Nuclear Science and Engineering was organized by Institut Teknologi Bandung. Then at the end of the session of SRs some topics about contribution of university to SRs R&Ds.

5.2. Breed-and-Burn Reactors (B&BRs)

Safety and economy are only two of the five problems inherent to nuclear energy, as shown in figure 2. We should be concerned with the other three problems: resources, waste, and nuclear threats.

Another important reactor design concept for the future is the B&BR. The standing wave reactor (SWR), traveling wave reactor (TWR) [20] and CANDLE reactor [7] pertain to this concept. These reactors use only natural uranium, depleted uranium, and/or thorium for their fresh fuel, and do not require reprocessing. Therefore, they are free from the nuclear threat problem. The burnup of these reactors is very high, i.e., 20% for SWR and 40% for TWR and CANDLE. Therefore, they can use fissile materials very efficiently, and the amount of spent fuel per energy produced is very small. These performances provide excellent solutions to resource and waste problems.

B&BRs have the excellent features above, but also difficult technical problems. One is the material problem caused by their very high burnup, and the other is the difficulty of criticality performance. R&D on these reactors began recently in the USA and basic studies have initiated in several emerging countries.

5.3. Contributions by Universities

The International Conference on Advances in Nuclear Science and Engineering was organized by Institut Teknologi Bandung. Then at the end of this Session of SRs some topics relating to universities are mentioned in the followings. In Japan the R&Ds of SRs are promoted by the international specialists' meeting SR/TIT organized by Tokyo Institute of Technology. The original idea of NuScale was born in Oregon State University, which is now running at the top of SRs development competition in the USA. And the concept of CANDLE was born in Tokyo Institute of Technology. Anyway, the nuclear energy was discovered by university professors. The contributions by universities to the innovative reactors should be large.

6. Roadmap

A technology roadmap of SRs and B&BRs is shown in figure. 6, together with conventional reactors.

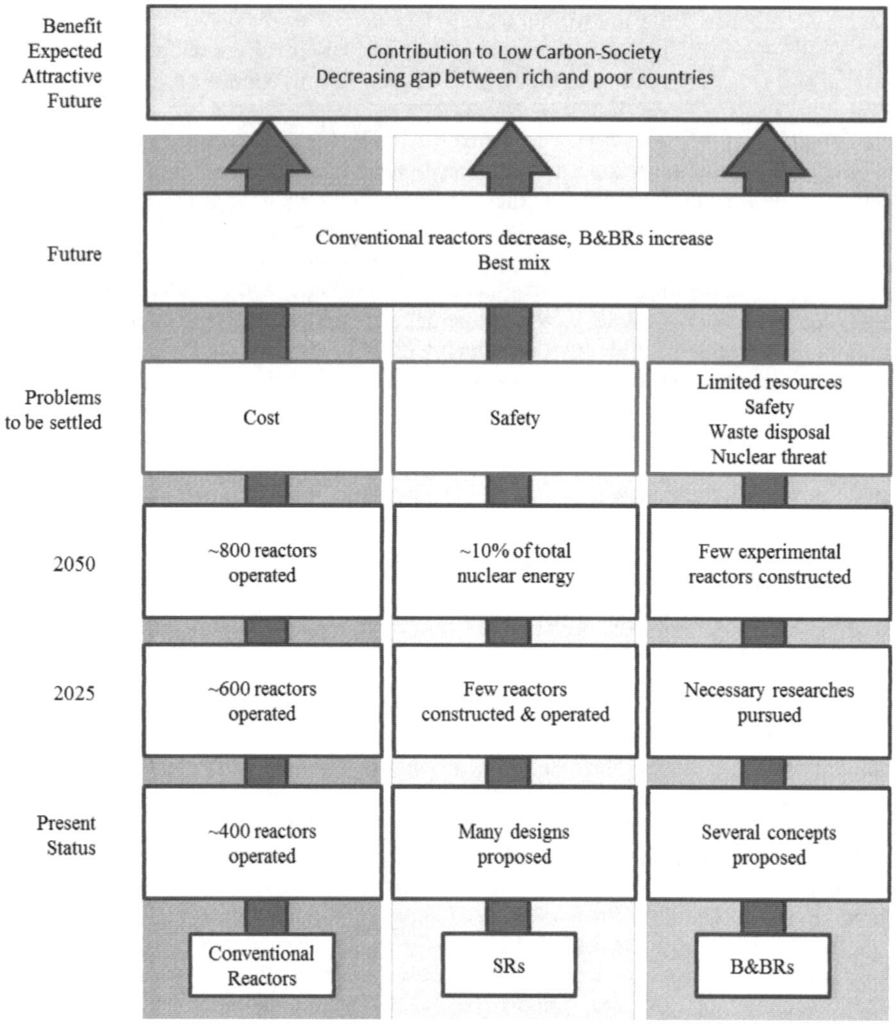

Figure 6. Technology roadmap of Japanese SRs and B&BRs, with predictions for conventional nuclear energy in Japan and forecasts of innovative nuclear energy in foreign countries (OECD countries, China, Russia and certain emerging countries).

The greatest obstruction to SRs is economics. However, we have many ideas to overcome this problem, including factory mass production [13]. SRs have many potential markets not accessible to large reactors. Further, they can be used in conventional large energy grids as modular reactors, demonstrating substantial advantages.

The major obstacle to B&BRs is technical, and is related to material development. It may take a long time to overcome this obstacle.

7. Benefit and Future Vision

Nuclear energy can considerably reduce greenhouse gas emissions and other environmental pollutants with acceptable cost. Daily control and maintenance of nuclear reactors are easily accomplished. However, as mentioned in Section 2, current nuclear energy has inherent and difficult problems to be solved at acceptable cost, i.e., limited resources, safety, waste disposal, and nuclear threats. The Fukushima accident has slowed nuclear reactor deployment, particularly in Japan. However, China [21] is aggressively promoting nuclear energy development. In many countries, these inherent problems are being studied more than before.

SRs will substantially solve the safety problem. B&BRs have great potential to solve all the inherent problems. SRs can realize special usages such as desalination, district heating, and certain industrial types. They can be used by ships and submarines, and are also expected to furnish undersea energy for investigation and mining. Many of these uses may be realized by the middle of this century. One additional important characteristic of nuclear energy is its very dense nature, as mentioned in Section 2. It needs no additional material like oxygen for fossil fuel. Therefore, it is almost the only useful energy for deep space, where in the far future, humans will attempt to penetrate. At that time, they will need nuclear energy. Nuclear reactors are currently producing considerable spent fuel waste. Most of this waste can be used as new nuclear fuel. Fissile concentration in the spent fuel of B&BRs is particularly high. Although presently only underground disposal is considered for nuclear waste, controlled storage should also be evaluated. In the future, transport to space may be much safer, cheaper and more reliable, if innovative techniques such as the space elevator [22] become available. At that time, storage of these wastes in artificial structures on/in the moon, asteroids, and outer space will become attractive options.

References

[1] Sekimoto H 2016 Nuclear Power Generation *Energy Technology Roadmaps of Japan*, Kato Y, et al. (eds.) (Springer, Japan) 257.
[2] International Atomic Energy Agency 2014 *Climate Change and Nuclear Power 2014*, http://www-pub.iaea.org/MTCD/Publications/PDF/ccanp2014web-14869824.pdf.
[3] Nuclear Energy Agency, International Energy Agency 2015 *Technology Roadmap - Nuclear Energy*, http://www.iea.org/publications/freepublications/publication/TechnologyRoadmap NuclearEnergy.pdf.
[4] Wikipedia 2014 *Energy density*, http://en.wikipedia.org/wiki/Energy_density.
[5] Sekimoto H 1990 *Nuclear Reactor Theory*, (COE-INES Textbook, Tokyo Institute of Technology, Tokyo, Japan).
[6] Nuclear Threat Initiative 2014 *Understanding Nuclear Threats*, http://www.nti.org/threats/nuclear/
[7] Sekimoto H 2010 *Light a CANDLE*, 2nd edition (CRINES, Tokyo Tech, ISBN978-4-905205-00-5 C3053 E, Japan).
[8] Sekimoto H 2014 A Brief History of Design Studies on Innovative Nuclear Reactors *Proc. of the 4th International Conference on Advances in Nuclear Science and Engineering, (ICANSE)* (Bali, Indonesia, September 16-19, 2013, AIP Proc., New York, USA).
[9] Wikipedia 2014 *Radioactive waste*, http://en.wikipedia.org/wiki/Radioactive_waste.
[10] Marcus GH 2010 *Nuclear Firsts: Milestones on the Road to Nuclear Power Development* (American Nuclear Society, Illinois, USA).
[11] Lilienthal D 1980 *Atomic Energy, a New Start* (Harper and Row, New York, USA).
[12] Weinberg AM, Spiewak I 1984 Inherently Safe Reactors and a Second Nuclear Era *Science* **224** 1398.
[13] Sekimoto H 1992 Several Features and Applications of Small Reactors *Potential of Small Nuclear Reactors for Future Clean and Safe Energy Sources* (Elsevier) 23.
[14] World Nuclear Association 2014 *Small Nuclear Power Reactors*, http://www.world-nuclear.org/info/Nuclear-Fuel-Cycle/Power-Reactors/Small-Nuclear-Power-Reactors.

[15] International Atomic Energy Agency 2012 *Status of Small and Medium Sized Reactor Designs*, http://www.iaea.org/books.
[16] International Atomic Energy Agency 2014 *Advances in Small Modular Reactor Technology Developments*, http://www.iaea.org/books.
[17] Wikipedia 2014 *Generation IV reactor*, http://en.wikipedia.org/wiki/Generation_IV_reactor.
[18] International Atomic Energy Agency 2014 *Technology Roadmap Update for Generation IV Nuclear Energy Systems*, www.gen-4.org/gif/upload/docs/application/pdf/2014-03/gif-tru2014.pdf.
[19] U.S. Department of Energy 2014 *FY 2014 Consolidated Innovative Nuclear Research FOA*, http://www.energy.gov/ne/downloads/fy-2014-consolidated-innovative-nuclear-research-foa-0.
[20] Gilleland J 2009 On the Traveling-Wave Reactor *Nuclear News* September (ANS, USA) 30.
[21] World Nuclear Association 2014 *Nuclear Power in China*, http://www.world-nuclear.org/info/Country-Profiles/Countries-A-F/China--Nuclear-Power/.
[22] Wikipedia 2014 *Space elevator*, http://en.wikipedia.org/wiki/Space_elevator.

IOP Conf. Series: Journal of Physics: Conf. Series **799** (2017) 012002 doi:10.1088/1742-6596/799/1/012002

A Simple Global View of Fuel Burnup

Hiroshi Sekimoto

Tokyo Institute of Technology

Email: hsekimot@gmail.com

Abstract. Reactor physics and fuel burnup are discussed in order to obtain a simple global view of the effects of nuclear reactor characteristics to fuel cycle system performance. It may provide some idea of free thinking and overall vision, though it is still a small part of nuclear energy system. At the beginning of this lecture, governing equations for nuclear reactors are presented. Since the set of these equations is so big and complicated, it is simplified by imposing some extreme conditions and the nuclear equilibrium equation is derived. Some features of future nuclear equilibrium state are obtained by solving this equation. The contribution of a nucleus charged into reactor core to the system performance indexes such as criticality is worth for understanding the importance of each nuclide. It is called nuclide importance and can be evaluated by using the equations adjoint to the nuclear equilibrium equation. Examples of some importances and their application to criticalily search problem are presented.

1. Introduction

"Free Thinking and Overall Vision" was an important guiding principle of the 21st Century Centre of Excellence (COE) Program "Innovative Nuclear Energy Systems for Sustainable Development of the World" promoted by Tokyo Institute of Technology where I was working as the leader [1]. This guiding principle is considered important not only for the Program but for all peoples involving in nuclear energy. Low operation rates, troubles and accidents met in nuclear power production activities may be attributed to the lack of it.

In the last September on Jeju Island, Korea, I made an invited lecture at the first Reactor Physics Asia Conference (RPHA), where I tried to make the attending reactor physics specialists get some idea of free thinking and overall vision. This lecture has been revised for the present lecture to be desirable for more general audiences.

The present lecture is on reactor physics and fuel burnup. It shows a simple global view of fuel burnup and reactor characteristics. I hope it provides us some ideas of free thinking and overall vision on this field, though it is still a small part of nuclear energy system.

At the beginning of this lecture, governing equations for nuclear reactors are presented. Since the set of these equations is so big and complicated, we try to simplify it by imposing some extreme conditions and to derive the nuclear equilibrium equation. Some features of future nuclear equilibrium state are obtained by solving this equation.

The contribution of each nucleus charged into a reactor core to its system performance index such as criticality is worth for understanding the importance of the nuclide. It is called nuclide importance and can be evaluated by using the equations adjoint to the nuclear equilibrium equation. Examples of some importances and their application to criticality search problems are presented in this lecture.

2. Basic Equations of Neutron Transport and Fuel Burning

2.1. Governing equations

For the investigation of global view of fuel burnup, the main parameters are neutron flux, $\phi(\vec{r},\vec{\Omega},E,t)$, and nuclide number density distributions, $n_j(\vec{r},t)$. In order to obtain these values we should treat the equations for the following distributions as governing equations:

(i) Neutron distribution
(ii) Delayed neutron precursor distribution
(iii) Nuclide (actinide, fission product (FP)) distribution
(iv) Temperature and material distribution

The actual equation for each distribution can be written as follows [2]:

(i) Neutron distribution

$$\begin{aligned}
&\frac{1}{v}\frac{\partial}{\partial t}\phi(\vec{r},\vec{\Omega},E,t)+\vec{\Omega}\cdot\nabla\phi(\vec{r},\vec{\Omega},E,t)+\sum_j n_j(\vec{r},t)\sigma_{T,j}(E)\phi(\vec{r},\vec{\Omega},E,t)\\
&\quad=\int_0^\infty dE'\int_{4\pi}d\vec{\Omega}'\sum_j n_j(\vec{r},t)\sigma_{S,j}(\vec{\Omega}'\cdot\vec{\Omega},E'\rightarrow E)\phi(\vec{r},\vec{\Omega}',E',t)\\
&\quad+\frac{1}{4\pi}\int_0^\infty dE'\int_{4\pi}d\vec{\Omega}'\sum_j n_j(\vec{r},t)\chi(E)\nu_j(1-\beta_j)\sigma_{F,j}(E')\phi(\vec{r},\vec{\Omega}',E',t)\\
&\quad+\sum_i\frac{\chi_i(E)}{4\pi}\lambda_i c_i(\vec{r},t)\\
&\quad+s(\vec{r},\vec{\Omega},E,t),
\end{aligned}\tag{1}$$

where

$\sigma_{T,j}(E)$: microscopic total cross section of nuclide *j* for the neutron with energy *E*,

$\sigma_{S,j}(\vec{\Omega}'\cdot\vec{\Omega},E'\rightarrow E)$: microscopic scattering cross section of the nuclide *j* for the neutron with energy *E*' and direction $\vec{\Omega}$' transferred to *E* and $\vec{\Omega}$. This reaction includes any neutron emitting reactions except fission,

$\sigma_{F,j}(E)$: microscopic fission cross section of nuclide *j* for the neutron with energy *E*,

ν_j : average number of neutrons emitted by the fission of nuclide *j*,

$\chi(E)$: prompt fission neutron energy spectrum,

$\chi_i(E)$: delayed neutron energy spectrum for precursor *i*,

β_j : delayed neutron precursor fraction produced by the fission of nuclide *j*,

λ_i : decay constant of delayed neutron precursor *i*,

$c_i(\vec{r},t)$: number density of delayed neutron precursor *i*,

$s(\vec{r},\vec{\Omega},E,t)$: external neutron source.

Though $\chi(E)\nu_j(1-\beta_j)$ is a function of the energy of impinging neutron which causes the fission, the expression (E') in the equation is omitted for simplicity.

(ii) Delayed neutron precursor distribution

$$\frac{\partial}{\partial t}c_i(\vec{r},t)+\lambda_i c_i(\vec{r},t)=\int_0^\infty dE'\int_{4\pi}d\vec{\Omega}'\sum_j n_j(\vec{r},t)\beta_{i,j}\nu_j\sigma_{F,j}(E')\phi(\vec{r},\vec{\Omega}',E',t), \tag{2}$$

where

$\beta_{i,j}$: fraction of delayed neutron precursor *i* produced by the fission of nuclide *j*, which satisfies

$$\beta_j=\sum_i\beta_{i,j}. \tag{3}$$

(iii) Nuclide (actinide, FP) distribution

$$\frac{\partial}{\partial t}n_j(\vec{r},t)=-n_j(\vec{r},t)\left(\lambda_j+\int_0^\infty dE'\sigma_{A,j}(E')\int_{4\pi}\phi(\vec{r},\vec{\Omega}',E',t)d\vec{\Omega}'\right)$$
$$+\sum_k n_k(\vec{r},t)\lambda_{k\to j}+\sum_k n_k(\vec{r},t)\int_0^\infty dE'\sigma_{k\to j}(E')\int_{4\pi}\phi(\vec{r},\vec{\Omega}',E',t)d\vec{\Omega}', \tag{4}$$

where

$\sigma_{A,j}(E)$: microscopic absorption cross section of nuclide *j* for the neutron with energy *E*,

$\sigma_{k\to j}(E)$: microscopic neutron cross section of nuclide *k* to be transformed to nuclide *j*,

λ_j : decay constant of nuclide *j*,

$\lambda_{k\to j}$: decay constant of nuclide *k* to be transformed to *j*.

Here the nuclide *n* covers both fuel and FPs. The production of FP from fission can also be treated by equation (4) by choosing $\sigma_{k\to j}(E)$ properly.

It should be noted that equation (4) does not include the in-core fuel managements. We need the initial conditions of nuclide number density distributions, $n_j(\vec{r},0)$, and/or in-core fuel management rules. About these points we will discuss in the next chapter for a very simple case.

(iv) Temperature and material distribution

In addition to the above equations we consider the equations of temperature and motion of materials. However, these equations depend on the core configuration very much. For example, we should use different equations for different coolant materials such as gas or liquid. Therefore, hereafter we omit to write the actual equations for these distribution. However, the coolant flow distribution affects strongly the temperature distributions, and the temperature distributions affect the effective microscopic cross sections and the material densities and shapes.

2.2. Steady state reactor operation without external source

In the following we consider the steady state reactor operation without external source, and the following equations hold:

$$\frac{1}{v}\frac{\partial}{\partial t}\phi(\vec{r},\vec{\Omega},E,t)=0, \tag{5}$$

$$\frac{\partial}{\partial t}c_j(\vec{r},t)=0, \tag{6}$$

$$s(\vec{r},\vec{\Omega},E,t)=0. \tag{7}$$

In this case the reactor is in a critical state and the external neutron source does not stay in the core. Only time dependent phenomena considered is the burnup.
In this case equations (1) and (2) can be combined to the following equation:

$$\begin{aligned}\vec{\Omega}\cdot\nabla\phi(\vec{r},\vec{\Omega},E,t)+\sum_j n_j(\vec{r},t)\sigma_{T,j}(E)\phi(\vec{r},\vec{\Omega},E,t)\\ =\int_0^\infty dE'\int_{4\pi}d\vec{\Omega}'\sum_j n_j(\vec{r},t)\sigma_{S,j}(\vec{\Omega}'\cdot\vec{\Omega},E'\rightarrow E)\phi(\vec{r},\vec{\Omega}',E',t)\\ +\frac{1}{4\pi k_{eff}(t)}\int_0^\infty dE'\int_{4\pi}d\vec{\Omega}'\sum_j n_j(\vec{r},t)\chi_j(E)\nu_j\sigma_{F,j}(E')\phi(\vec{r},\vec{\Omega}',E',t).\end{aligned} \tag{8}$$

The reactor is usually kept critical by adjusting the control rods. However, it is complicated to implement this procedure in neutron transport equation, and the effective neutron multiplication factor, $k_{eff}(t)$, is introduced instead. The reactor is controlled to produce its total power output to be equal to the planned value, $P(t)$:

$$\Gamma\int_V d\vec{r}\int_0^\infty dE'\int_{4\pi}d\Omega'\sum_j n_j(\vec{r},t)\sigma_{F,j}(E')\phi(\vec{r},\Omega',E',t)=P(t), \tag{9}$$

where $\int_V \bullet\, d\vec{r}$ denotes the integration over the whole core.
The basic equations for our study are equations (4), (8) and (9) and the equations for temperature and material distribution. In the following discussions the equations for temperature and material distribution are omitted from the list of equations.

2.3. *Homogeneous infinite medium*

In the present paper we study on the homogeneous infinite medium, and the following equations hold:

$$k_{eff}(t)\rightarrow k_\infty(t), \tag{10}$$

$$P(t)\rightarrow p(t). \tag{11}$$

where $k_\infty(t)$ is the infinite medium neutron multiplication factor and $p(t)$ is the power density. In this case the arguments $\vec{r}$ and $\vec{\Omega}$ disappear and

$$\vec{\Omega}\cdot\nabla\phi(\vec{r},\vec{\Omega},E,t)=0, \tag{12}$$

and the following new variable is introduced:.

$$\phi(E,t)\equiv\int\phi(\vec{r},\vec{\Omega},E,t)d\vec{\Omega}. \tag{13}$$

The governing equations become

$$\sum_j n_j(t)\sigma_{T,j}(E)\phi(E,t)-\sum_j n_j(t)\int_0^\infty\sigma_{S,j}(E'\rightarrow E)\phi(E',t)dE'$$

$$= \frac{1}{k_\infty(t)} \sum_j n_j(t)\chi_j(E) \int_0^\infty \nu_j \sigma_{F,j}(E')\phi(E',t)dE', \tag{14}$$

$$\Gamma \sum_j n_j(t) \int_0^\infty \sigma_{F,j}(E')\phi(E',t)dE' = p(t), \tag{15}$$

$$\frac{\partial}{\partial t} n_j(t) = -n_j(t)\left(\lambda_j + \int_0^\infty \sigma_{A,j}(E')\phi(E',t)dE' \right) + \sum_k n_j(t)\lambda_{k\to j} + \sum_k n_k(t) \int_0^\infty \sigma_{k\to j}(E')\phi(E',t)dE'. \tag{16}$$

2.4. Integration over neutron energy

For the study of fuel burnup, the variable E in neutron flux is not important if we can get one-group cross section set. Here we try to integrate equations (14) through (16). At the beginning we define the following one-group constants:

$$\phi(t) \equiv \int \phi(E,t)dE, \tag{17}$$

$$\sigma_{A,j} \approx \sigma_{A,j}(t) \equiv \frac{\int \sigma_{A,j}(E)\phi(E,t)dE}{\int \phi(E,t)dE}, \tag{18}$$

$$\sigma_{F,j} \approx \sigma_{F,j}(t) \equiv \frac{\int \sigma_{F,j}(E)\phi(E,t)dE}{\int \phi(E,t)dE}, \tag{19}$$

$$\sigma_{k\to j} \approx \sigma_{k\to j}(t) \equiv \frac{\int \sigma_{k\to j}(E)\phi(E,t)dE}{\int \phi(E,t)dE}. \tag{20}$$

Then equations (14) through (16) can be rewritten as follows:

$$\sum_j n_j(t)\sigma_{A,j}\phi(t) = \frac{1}{k_\infty(t)} \sum_j n_j(t)\nu_j \sigma_{F,j}\phi(t), \tag{21}$$

$$\Gamma \sum_j n_j(t)\sigma_{F,j}\phi(t) = p(t), \tag{22}$$

$$\frac{\partial}{\partial t} n_j(t) = -n_j(t)\left(\lambda_j + \sigma_{A,j}\phi(t)\right) + \sum_k n_k(t)\lambda_{k\to j} + \sum_k n_k(t)\sigma_{k\to j}\phi(t). \tag{23}$$

Equations (21) and (22) will be rewritten further as follows:

$$k_\infty(t) = \frac{\sum_j n_j(t)\nu_j \sigma_{F,j}}{\sum_j n_j(t)\sigma_{A,j}}, \tag{24}$$

$$\phi(t) = \frac{p(t)}{\Gamma \sum_j n_j(t)\sigma_{F,j}}. \tag{25}$$

Here we should be careful at the denominator of equation (24), neutron absorption term. The suffix j covers all of the nuclides in the core. However, in other parts j covers only nuclides produced in the core from the fuel, which are actinides and FPs. Now we define h-value as follows:

$$h(t) = \frac{\sum_{j \in Act} n_j(t)\nu_j \sigma_{F,j}}{\sum_{j \in Act, FP} n_j(t)\sigma_{A,j}}. \tag{26}$$

And h-value without FP as follows:

$$h_{Act}(t) = \frac{\sum_{j \in Act} n_j(t)\nu_j \sigma_{F,j}}{\sum_{j \in Act} n_j(t)\sigma_{A,j}}. \tag{27}$$

Apparently the following equation holds:

$$h_{Act}(t) > h(t) > k_\infty(t). \tag{28}$$

Now we have delivered all necessary equations for our study, equations (23), (24) and (25). These equations consist the governing equations together with the equations for temperature and material distribution.

3. Fuel Burning Equation at Continuous Operation

3.1. Fuel burning equation at continuous operation

In the following parts equation (23) will appear often, and then we will rewrite it by using slightly different notations for convenience as follows:

$$\frac{dn_j}{dt} = -(\lambda_j + \sigma_{A,j}\phi)n_j + \sum_k (\alpha_{k\to j}\lambda_k + \beta_{k\to j}\sigma_{A,k}\phi + \gamma_{k\to j}\sigma_{F,k}\phi)n_k , \tag{29}$$

where

λ_j : decay constant of nuclide j,

$\sigma_{A,j}$: microscopic absorption cross-section of nuclide j,

$\sigma_{F,j}$: microscopic fission cross-section of nuclide j,

$\alpha_{k\to j}$: branching ratio of decay of nuclide k to nuclide j,

$\beta_{k\to j}$: branching ratio of neutron absorption of nuclide k to nuclide j,

$\gamma_{k\to j}$: branching ratio of fission of nuclide k to nuclide j.

Equatiton (26) does not include fuel managements. We need it for solving this equation. Here we study on continuous operation, where fuel charge and discharge are continuous. In this case we can treat the problem simply by adding charge and discharge terms into equation (26) as follows:

$$\frac{dn_j}{dt} = -\left(\lambda_j + \sigma_{A,j}\phi + r_j\right)n_j + \sum_k \left(\alpha_{k\to j}\lambda_k + \beta_{k\to j}\sigma_{A,k}\phi + \gamma_{k\to j}\sigma_{F,k}\phi\right)n_k + s_j, \tag{30}$$

where

r_j : discharge constant of nuclide j,
s_j : charge rate of nuclide j.

This equation can be written in a matrix form as follows:

$$\frac{d\boldsymbol{n}}{dt} = -\boldsymbol{M}\boldsymbol{n} + \boldsymbol{s}, \tag{31}$$

where

$\boldsymbol{M}$: transmutation matrix,
$\boldsymbol{n}$: nuclide densities of actinides and FPs ,
$\boldsymbol{s}$: fuel supply rate.

The matrix $\boldsymbol{M}$ and vector $\boldsymbol{n}$ can be written as

$$\boldsymbol{M} = \begin{bmatrix} \boldsymbol{M}_{Act,Act} & \boldsymbol{0} \\ -\boldsymbol{M}_{Act,FP} & \boldsymbol{M}_{FP,FP} \end{bmatrix}, \tag{32}$$

$$\boldsymbol{n} = \begin{bmatrix} \boldsymbol{n}_{Act} \\ \boldsymbol{n}_{FP} \end{bmatrix}. \tag{33}$$

Where suffixes Act and FP represent actinide and FP, respectively, and equation (31) can be discomposed as

$$\frac{d\boldsymbol{n}_{Act}}{dt} = -\boldsymbol{M}_{Act,Act}\boldsymbol{n}_{Act} + \boldsymbol{s}_{Act}, \tag{34}$$

$$\frac{d\boldsymbol{n}_{FP}}{dt} = \boldsymbol{M}_{Act,FP}\boldsymbol{n}_{Act} - \boldsymbol{M}_{FP,FP}\boldsymbol{n}_{FP} + \boldsymbol{s}_{FP}. \tag{35}$$

From these equations we can see that the actinide density, $\boldsymbol{n}_{Act}$, can be obtained without any information about FPs, and h_{Act}-value can be obtained by solving only equation (34).
The conventional reactor operation is batch operation, where the fuel charge and discharge operations are performed during the reactor shutdown period. In this reactor operation scheme $\boldsymbol{s} = 0$ during the reactor operation period, and (31) can be solved as

$$\boldsymbol{n} = e^{-\boldsymbol{M}t}\boldsymbol{n}_B, \tag{36}$$

where $\boldsymbol{n}_B$ is the nuclide density vector at the beginning of cycle (BOC). For the cycle length of T, the nuclide density vector at the end of cycle (EOC) $\boldsymbol{n}_E$ is given by

$$\boldsymbol{n}_E = e^{-\boldsymbol{M}T}\boldsymbol{n}_B. \tag{37}$$

If $\boldsymbol{q}$ is charged and $\boldsymbol{d}$ is discharged during the shutdown period, $\boldsymbol{n}_B$ can be given by the following equation:

$$\boldsymbol{n}_B = \boldsymbol{n}_E + \boldsymbol{q} - \boldsymbol{d}. \tag{38}$$

By substituting Eq. (38) into equation (37)

$$\boldsymbol{n}_E = e^{-\boldsymbol{M}T}(\boldsymbol{n}_E + \boldsymbol{q} - \boldsymbol{d}).$$

Then

$$(e^{\boldsymbol{M}T} - \boldsymbol{I})\boldsymbol{n}_E = (\boldsymbol{q} - \boldsymbol{d})/T.$$

When $\|\boldsymbol{M}T\| << 1$

$$\boldsymbol{M}\boldsymbol{n}_E = (\boldsymbol{q} - \boldsymbol{d})/T. \tag{39}$$

Usually we treat the set of equations (34) and (35) instead of equation (31).
We can rewrite h and h_{Act} by using vector notations as follows:

$$h = \frac{(\nu\sigma_f, \boldsymbol{n}_{Act})}{(\sigma_{a,}\boldsymbol{n})}, \tag{40}$$

$$h_{Act} = \frac{(\nu\sigma_f, \boldsymbol{n}_{Act})}{(\sigma_{a,}\boldsymbol{n}_{Act})}. \tag{41}$$

3.2. *Nuclear equilibrium state*

From the finiteness of the earth our society ultimately should become a steady state society where the energy consumption rate is a constant [3]. If this energy supply is satisfied by the fission energy, the fresh fuel charge rate and waste discharge rate are also constants. Furthermore, the reaction rate of each nuclear reaction and abundance ratio of each nuclide in the reactor are also constants [4]. This kind of state is called as the nuclear equilibrium state.

I call the society in this state the nuclear equilibrium society and continues to work under this research title for long period. In this society the energy production system is considered to affect the social system strongly. It is interested to know what kind of society will be realized as such a future society. However, we have many scientific and engineering problems to be investigated before studying the society itself.

We consider a nuclear energy center as shown in figure 1, in which several nuclear reactors and handling and storage facilities of nuclear fuels and wastes are operated. In the present scenario any radioactive materials will not be taken out of the nuclear energy center.
Although we can consider many kinds of the nuclear equilibrium systems, we will study on simple systems at the beginning. The number of types of reactor is fixed to be one in order to know the fundamental effects of the reactor to the system, though the system with several kinds of reactors are possible and may work as an efficient synergetic system. The number of type of fuel is also chosen to be one, natural uranium or thorium.

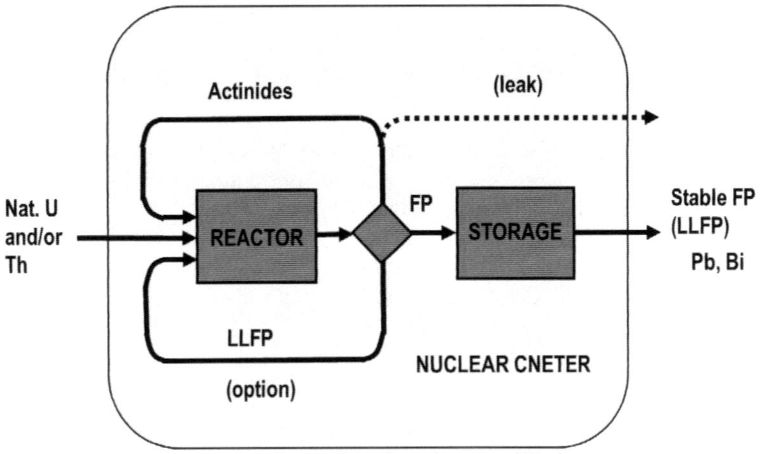

Figure 1. A simple model of nuclear equilibrium system

In the present model, the natural uranium or thorium are supplied to the reactor in the center as shown in figure 1. All the actinides (including the daughter nuclides in the secular equilibrium with this nuclide) are confined in the reactor. Almost all of them will fission and only small amount becomes the terminal nuclides after radioactive decays; ^{204}Pb, ^{206}Pb, ^{207}Pb, ^{208}Pb and ^{209}Bi. In this system these terminal nuclides of actinide decay chain and FPs are withdrawn from the reactor. Otherwise they are accumulated in the reactor.

3.3. Nuclear equilibrium equation

The nuclide composition in the reactor does not change along time for the nuclear equilibrium state. The rigid nuclear equilibrium state is investigated for the continuous operation as the simplest case. It may be a good approximation even for the batch-wise discontinuous fuel-cycle operation, and useful for the investigation of change of the state characteristics for different neutron spectra and inserted fuel.
Now we consider a region of unit volume in the reactor. Fuel is considered to be charged to this region with a constant rate. The nuclide that does not relate to the natural uranium or thorium is not considered in this analysis. The nuclides are transmuted in the reactor to other nuclides. The transmutation is performed with a nuclear reaction with a neutron and natural decays. Through these transmutations many kinds of nuclides present in the reactor in the equilibrium state.
The strict equilibrium nuclear state is a steady state, where the nuclide number density of each nuclide does not change, and the following equation holds:

$$\frac{d\boldsymbol{n}}{dt} = 0. \qquad (42)$$

Therefore, Eq. (31) becomes

$$\boldsymbol{Mn} = \boldsymbol{s}\,. \tag{43}$$

or equations (34) and (35) become

$$\boldsymbol{M}_{Act,Act}\boldsymbol{n}_{Act} = \boldsymbol{s}_{Act}, \tag{44}$$

$$\boldsymbol{M}_{FP,FP}\boldsymbol{n}_{FP} = \boldsymbol{M}_{Act,FP}\boldsymbol{n}_{Act} + \boldsymbol{s}_{FP}\,. \tag{45}$$

This equation is called as the nuclear equilibrium equation.
It is noted that the nuclear equilibrium equation (43) is similar to equation (39) which holds for $\|\boldsymbol{MT}\| << 1$.
The nuclear equilibrium equation is a linear algebraic equation, and can be easily solved for constant coefficients. Many of these coefficients are the one-group constants defined by equations (18) through (20), which are neutron spectrum weighted values. When the neutron spectrum is given, these coefficients are constants and the nuclear equilibrium equation can be solved very easily. However, neutron spectrum changes for different nuclide density configurations. In such a case, ***M*** depends on ***n***. and we need a coupling code system of neutron spectrum calculation and nuclear equilibrium calculation. One of such a code system is shown in figure 2 [5].

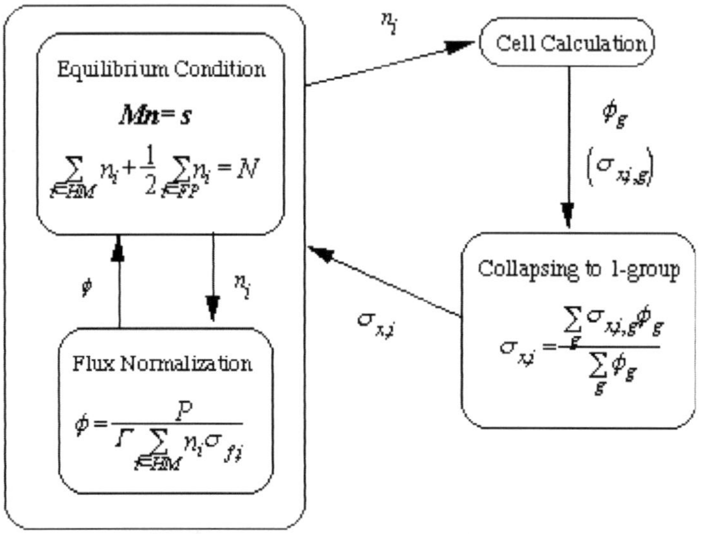

Figure 2. Equilibrium Cell Iterative Calculation System (ECICS) [5]

3.4. Solution for Some Simple Problems [4]
Here we will solve equation (43) and obtain h-value by using equation (40) for a simple problem in order to investigate the possibility to realize a typical nuclear equilibrium state [4].
3.4.1. Calculation Conditions
Equation (43) is so simple that we can treat many nuclides without any calculation problems. All actinides whose half-life is more than one day are employed in the present analysis. Then 129 actinides and 1238 FPs are treated in the present study. In usual cases, however, much less number of nuclides is enough for evaluating neutron economy in an acceptable accuracy.

As typical reactors two types of fast reactors and two types of thermal reactors are investigated. For the fast reactors a sodium-cooled oxide-fuel fast reactor (soft fast) and a sodium-cooled metallic fuel fast reactor (hard fast) are treated, and for the thermal reactors a PWR (hard thermal) and a high moderation-ratio graphite-moderated gas-cooled reactor (soft thermal) are treated. The neutron spectra of these reactors are shown in figure 3.

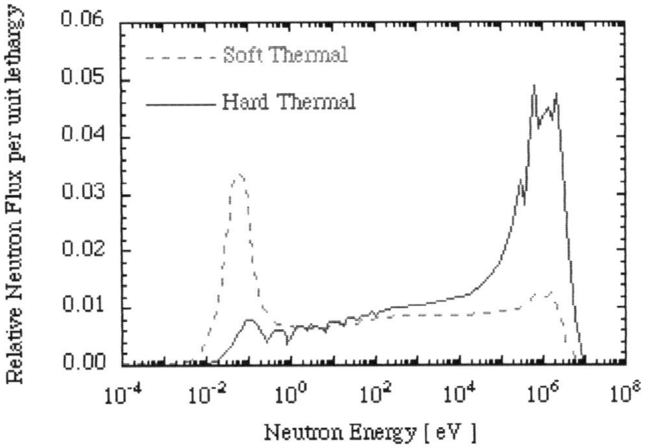

(a) Thermal reactors

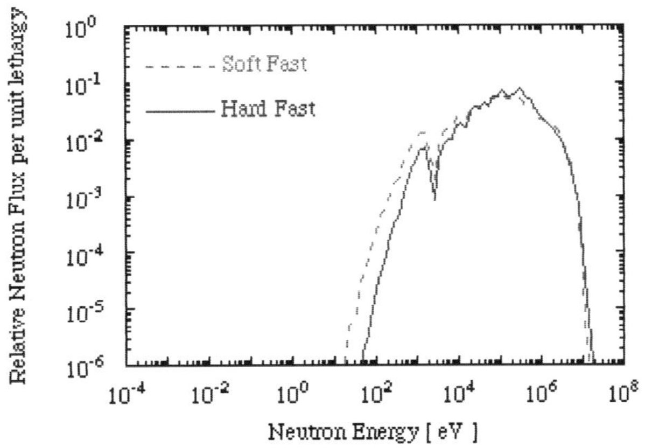

(b) Fast reactors

Figure 3. Typical neutron spectra

The one-group constants are obtained by using these spectra and equations (18) thorough (20). Fundamental design parameters of the reactors are shown in table 1. The removal half-life is given by $T_{1/2} = \ln 2/r$, where r is the discharge constant (See equation (30).). These parameters are enough for the present stage of the study, since only general aspects are interested in.

Table 1. Fundamental reactor design parameters

	Thermal Reactor	Fast Reactor
Total Power Output	3 GWt	3 GWt
Power Density	50 W/cc	300 W/cc
Removal Half Life	1 year	1 year

3.4.2. Nuclide Densities

The obtained nuclide densities of all actinides more than $10^{15}/cm^3$ in a reactor charged with natural uranium or thorium are shown in figure 4 for soft fast or hard thermal reactors. Though in these figures only the cases for soft fast and hard thermal reactors are shown, the differences between hard fast and soft fast and between hard thermal and soft thermal are much smaller than the differences between fast and thermal.

The actinide nuclide number densities do not depend on the FP discharge constant, since they are obtained by solving (44) which does not include the FP discharge constant.

Apparently the densities of higher actinides are higher for the natural-uranium charge case than the thorium charge case and vice versa. The densities of charged nuclides are about the same for all four cases. The decrease amount from the charged nuclide to the next nuclide is larger for the thermal reactors than the fast reactors but the decreasing slope becomes slower.

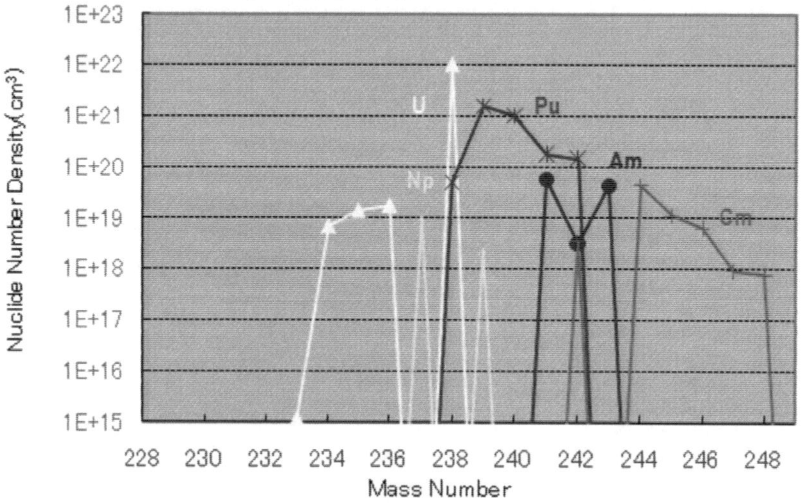

(a) Soft fast reactor charged with natural uranium

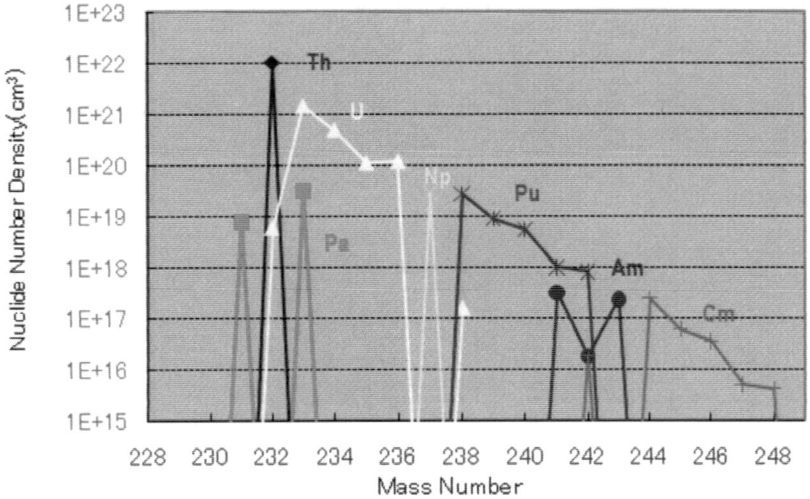

(b) Soft fast reactor charged with thorium

Figure 4. Equilibrium nuclide densities for different nuclear systems [5]

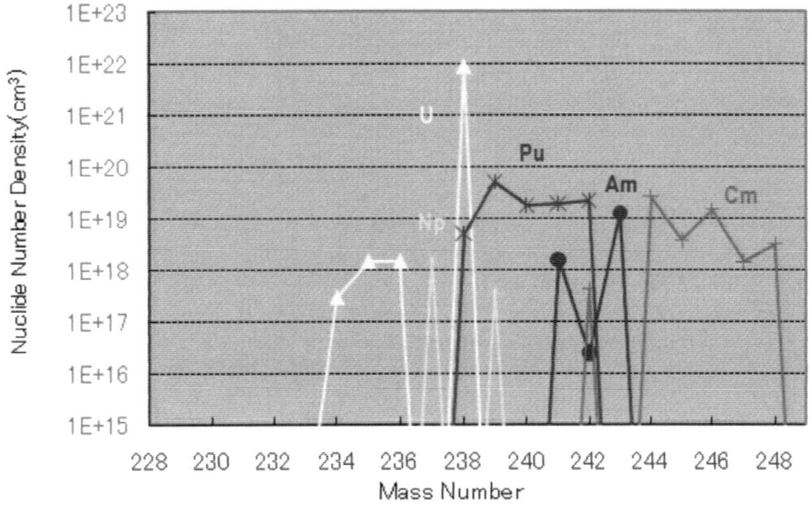

(c) Hard thermal reactor charged with natural uranium

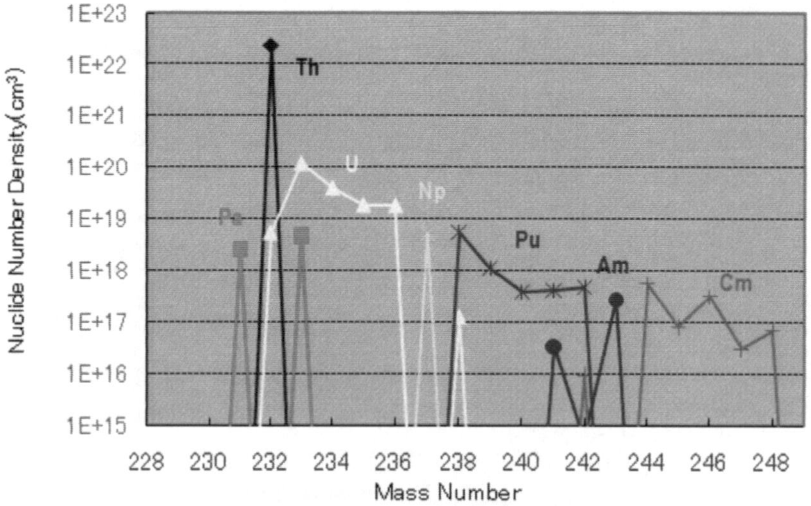

(d) Hard thermal reactor charged with thorium

Figure 4. (continued)

From these nuclide densities we can obtained many characteristics of each system. In the next section we will discuss the criticality. Discharged fuel characteristics are also easily evaluated by using these nuclide densities. However these topics are omitted from the page limit. I hope the readers who are interested in these topics will see some of the references [6, 7].

3.4.3. Criticality

The h-values obtained using the previously shown nuclide densities are shown in figure 5 for the FP removal half-life of 1 year, where the power density level is changed to investigate its effect. The horizontal axis denotes the neutron flux level which is proportional to the power density level. The flux level change causes change of the ratio of the neutron induced reaction to radioactive decay. Therefore, the nuclide density distribution along mass number will change, and h-value changes. For the thorium cycle case, h-value decreases with increasing flux level. It is attributed to the increasing neutron absorption by ^{233}Pa for increasing flux level.

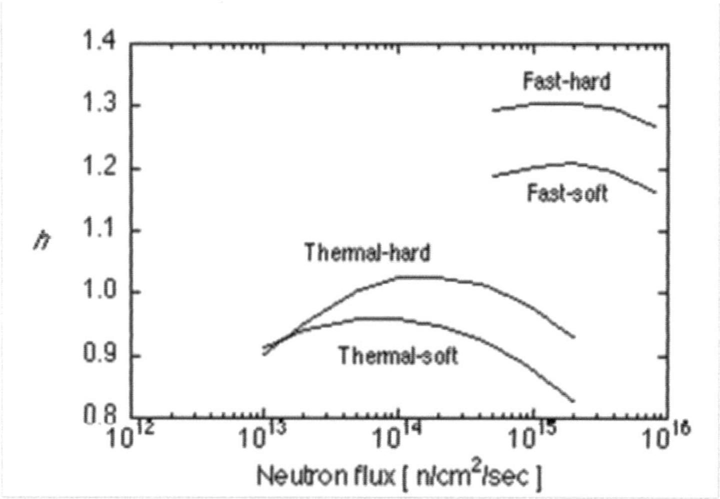

(a) Natural uranium feed case

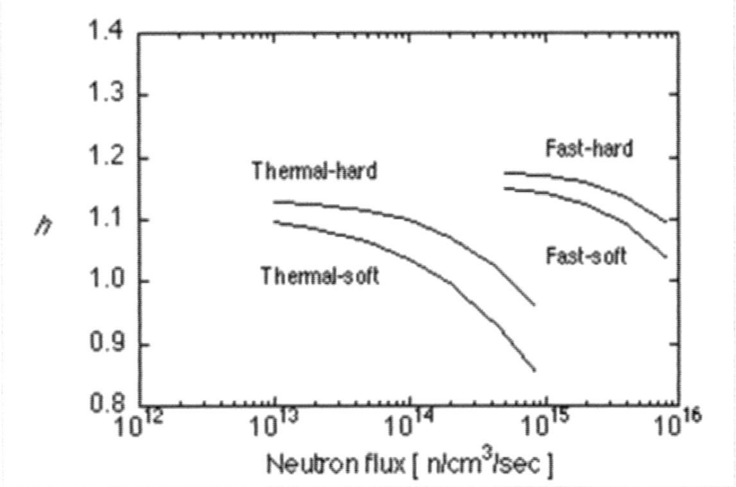

(b) Thorium feed case

Figure 5. h-values for different nuclear system [5]

When we consider the neutron economy, 5 - 10 % of produced neutrons are usually absorbed by construction materials, coolant, moderator, control materials, etc., and leakage from the system, though this amount depends strongly on the reactor design. Therefore, the h-value should take a value more than 1.03 - 1.1 to perform the criticality. Judging from these calculation results, the fast reactor charged with natural uranium can perform the criticality with some surplus, but the thermal reactor charged with natural uranium cannot perform it. The system with thorium may perform the criticality for both fast and thermal neutron system for lower neutron flux level, but it seems not easy. The fast reactor system shows the better performance also for this system, but the difference is small. The system showing the best neutron economy is the fast reactor system charged with natural uranium.

4. Nuclide Importances [8]

4.1. Importance and adjoint function

At the beginning on this chapter I will mention about some basic theory of linear systems. We start the discussion with the following linear state equation (governing equation)

$$M|n\rangle = |s\rangle . \tag{46}$$

This equation is a simple rewriting of equation (43) by using bra-vector and ket-vector, which are the notations introduced by Dirac for describing quantum mechanics [9], but convenient for describing our equations. The ket-vector $|n\rangle$ is called the state vector, and $|s\rangle$ is called the source vector. We consider an objective value (performance index) which is given by a scalar product of the source vector and the weight vector $\langle w|$ as follows:

$$Q = \langle w|n\rangle . \tag{47}$$

If this objective value can be written as

$$Q = \langle i | s \rangle . \tag{48}$$

The bra-vector $\langle i|$ is called an importance vector, or i_j is called an importance of s_j.
From equations (46) through (48)

$$\langle w | n \rangle = Q = \langle i | s \rangle = \langle i | M | n \rangle .$$

Therefore

$$\langle i | M = \langle w | . \tag{49}$$

This is the adjoint equation to equation (46), where the adjoint source is the weight vector. We see the importance vector is the adjoint vector.
Equations (44) and (45) can be rewritten by the present way

$$M_{Act,Act} | n_{Act} \rangle = | s_{Act} \rangle , \tag{50}$$

$$M_{FP,FP} | n_{FP} \rangle = M_{Act,FP} | n_{Act} \rangle + | s_{FP} \rangle . \tag{51}$$

The equations corresponding to equations (47) and (48) are as follows:

$$Q = \langle w_{Act} | n_{Act} \rangle + \langle w_{FP} | n_{FP} \rangle , \tag{52}$$

$$Q = \langle i_{Act} | s_{Act} \rangle + \langle i_{FP} | s_{FP} \rangle . \tag{53}$$

And the adjont equations are as follows:

$$\langle i_{Act} | M_{Act,Act} = \langle i_{FP} | M_{Act,FP} + \langle w_{Act} | , \tag{54}$$

$$\langle i_{FP} | M_{FP,FP} = \langle w_{FP} | . \tag{55}$$

4.2. Examples of importance
Some examples of weight (adjoint source) and corresponding importance are shown in table 2.

Table 2. Adjoint sources, w_j, for several importances, i_j, of nuclide j [8]

	i_j	w_j
nuclide number importance (total number of nuclides)	$i_{n,j}$	1
fission number importance (total number of fissions)	$i_{f,j}$	$\phi\sigma_{f,j}$
fission neutron importance (total number of produced fission neutrons)	$i_{\nu f,j}$	$\phi\nu\sigma_{f,j}$
absorbed neutron importance (total number of absorbed neutrons modified by neutron producing reactions except fission)	$i_{a,j}$	$\phi\left(\sigma_j - \sum_m m\sigma_{m,j}\right)$
net neutron production importance (net number of produced neutrons)	$i_{P,j}$	$\phi\left(\nu\sigma_{f,j} - \sigma_j + \sum_m m\sigma_{m,j}\right)$

In this section some calculated values of importances are presented for the systems shown in section *3.4*, but the neutron spectra of fission neutron, soft fast reactor (fast reactor), hard thermal reactor (thermal reactor) and thermal Maxwellian spectrum are used to obtain one-group cross sections for studying the difference of neutron spectrum.

The first example is the fission number importance, which is shown in table 3. The importances of important actinides are almost unity for reactor spectra. It means that almost all important actinides will ultimately fission for these conditions. Where the neutron energy becomes higher, the number of nuclides produced by (n, 2n) reaction of light actinide increases, and the some small part of produced nuclides escape fission and decay more easily to lighter nuclides finally to a stable isotope such as ^{204}Pb, ^{206}Pb, ^{207}Pb, ^{208}Pb and ^{209}Bi as mentioned in section *3.2*. And then the fission number importance decreases from unity by a small amount. The lower neutron flux level makes the importance decrease, since the ratio of fission reaction to radioactive decay decreases.

The next example is the net neutron importance, which is shown in figure 6. The results in this figure are for the case of FP discharge constant of infinity and 1/2 y^{-1} and for the case of neutron flux level of 10^{15} $cm^{-2}s^{-1}$. The importance generally increases with increasing mass number. This behavior is monotonic for the fission neutron spectra, but it is oscillatory for the reactor spectra. For the change of discharge constant from infinity to 1/2 y^{-1}, the importance shifts downward. This tendency is especially dominant for the softer spectrum. For the fast reactor, the importances of almost all actinides are positive for all cases, but the ^{236}U importance is negative. The negativeness becomes more severe for the thermal reactor. Then for the thermal reactor the multi-recycle of uranium makes neutron economy considerably worse.

More importances and more detail discussions on these results are given in reference [8].

Table 3. Fission number importance [8]

Neutron Spectrum	fission		fast reactor	thermal reactor	Maxwellian	
Flux Level ($cm^{-2}sec^{-1}$)	10^{15}	10^{16}	$4.72x10^{15}$	$2.17x10^{14}$	10^{15}	10^{16}
Th-232	0.99726	0.99972	0.99994	0.99989	1.00000	1.00000
U-233	0.99974	0.99997	1.00000	1.00000	1.00000	1.00000
U-234	1.00000	1.00000	1.00000	1.00000	1.00000	1.00000
U-235	1.00000	1.00000	1.00000	1.00000	1.00000	1.00000
U-236	1.00000	1.00000	1.00000	1.00000	1.00000	1.00000
U-238	1.00000	1.00000	1.00000	1.00000	1.00000	1.00000
Np-237	1.00000	1.00000	1.00000	1.00000	1.00000	1.00000
Pu-238	1.00000	1.00000	1.00000	1.00000	1.00000	1.00000
Pu-239	1.00000	1.00000	1.00000	1.00000	1.00000	1.00000
Pu-240	1.00000	1.00000	1.00000	1.00000	1.00000	1.00000
Pu-241	1.00000	1.00000	1.00000	1.00000	1.00000	1.00000
Pu-242	1.00000	1.00000	1.00000	1.00000	1.00000	1.00000
Am-241	1.00000	1.00000	1.00000	1.00000	1.00000	1.00000
Am-242m	1.00000	1.00000	1.00000	1.00000	1.00000	1.00000
Am-243	1.00000	1.00000	1.00000	1.00000	1.00000	1.00000
Cm-243	1.00000	1.00000	1.00000	1.00000	1.00000	1.00000
Cm-244	1.00000	1.00000	1.00000	1.00000	1.00000	1.00000
Cm-245	1.00000	1.00000	1.00000	1.00000	1.00000	1.00000
Cm-246	1.00000	1.00000	1.00000	1.00000	1.00000	1.00000
Cm-247	1.00000	1.00000	1.00000	1.00000	1.00000	1.00000
Cm-248	1.00000	1.00000	1.00000	1.00000	1.00000	1.00000

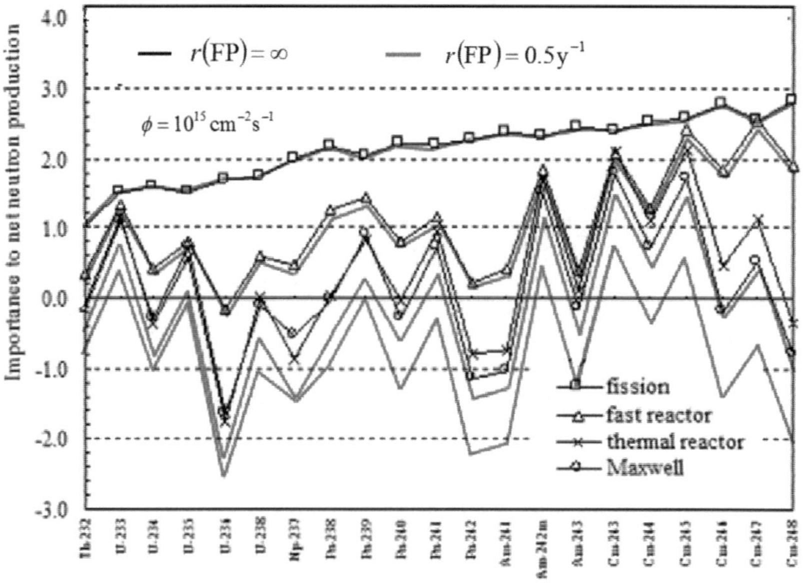

Figure 6. Net neutron importance ($\phi = 10^{15}\mathbf{cm}^{-2}\mathbf{s}^{-1}$) [8]

4.3. Criticality search for continuous operation

Generally criticality search is complicated and difficult to be performed. Even the simple system given in chapter 3 may take a lot of time for criticality search by adjusting flesh fuel compositions. If we employ the nuclide importances, it becomes a simple and easy procedure [10]. Here we try to find the charged fuel compositions to make a system critical for equilibrium cycle.

For the criticality condition, $k_{eff} = 1$, we consider that h-value takes the value h_c, where

$$h_c > 1. \tag{56}$$

If we have an acceptably accurate value of h_c, we can obtain the proper source term $|s\rangle$ to give the criticality by solving the following equation:

$$\langle i_{\nu f} - h_c i_a | s \rangle = 0, \tag{57}$$

where the importances $\langle i_{\nu f}|$ and $\langle i_a|$ are obtained from the following equations:

$$\langle i_{\nu f}| M = \langle \phi \nu \sigma_f |, \tag{58}$$

$$\langle i_a | M = \left\langle \phi \left(\sigma - \sum_m m\sigma_m \right) \right| \quad . \tag{59}$$

We will show some examples for demonstration of this methods, where the reactor is a PWR and its design parameters are shown in table 4 [10].

Table 4. Design parameters of studied PWR

Power Output (MW thermal)	3000
Average power density of Pellet (Wcm^{-3})	280
Average power density of Cell (Wcm^{-3})	100
Fuel Pellet diameter (mm)	8
Pin diameter (mm)	9.6
Pin pitch (mm)	11.8
Materials	
Fuel pellet	Oxide
Cladding	Zircaloy 4
Coolant	Light water
Volume fraction (%)	
Fuel pellet	36
Cladding	16
Coolant	48

We apply five in-core fuel management cases as shown in table 5, and try to find the enrichment and some other reactor characteristics for each scheme. The discharge constant for the standard operation is chosen as 1/3 yr^{-1}, in order to simulate in-core fuel management for three batches. The calculation system like figure 2 should be employed, since the neutron spectrum changes for different cases, and the changes of parasitic absorptions by non-fuel materials (construction materials and water) and leakage should be also evaluated. The details of this procedure are written in reference [10]

Table 5. Discharge constant for 5 in-core fuel managements ($r = 1/3\ yr^{-1}$)

Case	U	Pu	Other Actinides
1	r	r	r
2	r	r/2	r
3	r	0	r
4	r	0	0
5	0	0	0

The obtained results are shown in table 6. The obtained results represent well the characteristics of once-through and recycle systems, which are discussed widely in nuclear energy society. The plutonium and minor actinides recycle can reduce the required natural uranium amount as expected, but their effects are not so large. If the enrichment is increased enough, we can burnup whole charged fuel and reduce the required natural uranium amount considerably. However, the value of enrichment becomes higher than 20 % which cannot be permitted for commercial uses, and the burnup of discharged fuel is too much for satisfying the material constraints.

Table.6. Required enrichment and several fuel cycle characteristics [10]

Case	h_c	Enrichment ($^w/_o$)	Charged fuel (t/y)	Flux ($cm^{-2}.\ s^{-1}$)	Burnup (GWd/t)	Required Natural U(t/y)	
						0.1 $^w/_0$ tail	0.3 $^w/_0$ tail
1	1.046	4.2	29	3.83×10^{14}	37.9	195.4	276.3
2	1.041	4	28.8	3.80×10^{14}	38.1	185.3	261.4
3	1.031	3.8	28.1	3.73×10^{14}	39	169.2	237.9
4	1.030	3.6	27.6	3.71×10^{14}	39.8	158.9	222.8
5	1.026	41.6	1.2	3.46×10^{14}	952.7	79.4	117.5

References

[1] Sekimoto H 2008 Progress in ''COE-INES'' *Progress in Nuclear Energy* **50** 71.
[2] Sekimoto H 2007 Part 2.1 Transport Equation and Diffusion Equation *Nuclear Reactor Theory* COE-INES, Tokyo Institute of Technology, Japan.
[3] Meadows D H 1972 *The Limit to Grouth* (New American Library).
[4] Sekimoto H and Takagi N 1991 Preliminary Study on Future Society in Nuclear Quasi-Equilibrium *J. Nucl. Sci. Technol.* **28** 941
[5] Mizutani A and Sekimoto H 1997 Calculational Method of One-Group Nuclear Constants in Nuclear Equilibrium State *J. Nucl. Sci. Technol.* **34** 596.
[6] Sekimoto H, Nakamura H and Takagi N 1996 Toxicity of Radioactive Waste Discharged from Nuclear Energy Center in the Future Equilibrium State *Ann. Nucl. Energy* **23** 663.
[7] Sekimoto H and Kanai K 1998 Trade-off between Neutron Excess and Long-Term Toxicity Reduction by Employing LLFP Incineration *Ann. Nucl. Energy* **25** 793.
[8] Sekimoto H and Nemoto A 2000 Nuclide Importance and the Steady-State Burnup Equation *Nucl. Sci. Engin.* **135** 84.
[9] Dirac P A M 1958 *The Principles of Quantum Mechanics* 4th edition (Oxford University Press).

[10] Waris A and Sekimoto H 2001 Characteristics of Several Equilibrium Fuel Cycles of PWR *J. Nucl. Sci. Technol.* **38** 517.

IOP Conf. Series: Journal of Physics: Conf. Series **799** (2017) 012003 doi:10.1088/1742-6596/799/1/012003

Accident Analysis Simulation in Modular 300MWt Gas Cooled Fast Reactor

Zaki Su'ud

[a]Nuclear and Biophysics Research Division, Faculty of Mathematics and Natural Sciences, Bandung Institute of Technology

*E-mail: szaki@fi.itb.ac.id

Abstract. Safety analysis of 300MWt helium gas cooled long-life fast reactors has been performed. The analysis of unprotected loss of flow(ULOF) and unprotected rod run-out transient overpower (UTOP) are discussed. Some simulations for 300 MWt He gas cooled fast reactors has been performed and the results show that the reactor can anticipate complete pumping failure inherently by reducing power through reactivity feedback and remove the rest of heat through natural circulations. GCFR relatively has hard spectrum so it has relatively small Doppler coefficient. In the UTOP accident case the analysis has been performed against external reactivity up to 0.002dk/k. In addition the steam generator design has also consider excess power during severe UTOP case..

1. Introduction

In post Fukushima nuclear accidents inherent safety capability is necessary against some hypothetical accidents such as unprotected loss of flow (ULOF), unprotected rod run-out transient over power (UTOP), unprotected loss of heat sink (ULOHS). Gas cooled fast reactors is one of the important candidate of 4th generation nuclear power plant and in this paper the safety analysis related to unprotected loss of flow in small long life gas cooled fast reactors has been performed. Accident analysis of unprotected loss of flow include coupled neutronic and thermal hydraulic analysis. Natural circulation based heat removal system is important to ensure inherent safety capability during unprotected accidents.

Therfore the system similar to RVACS (reactor vessel auxiliary cooling system) is investigated. As the results some simulations for this 300 MWt gas cooled fast reactors has been performed and the results show that the reactor can anticipate complete pumping failure inherently by reducing power through reactivity feedback and remove the rest of heat through natural circulations.

2. Methodology[1,6-14]

The detail model and mathematical formulations for the current safety analysis can be obtained in the following references. Here the algorithm of the simulation will be discussed. After calculating effective macroscopic group constant then steady state multi-group diffusion calculation and steady state thermal hydraulic calculation are performed. The accident simulation begins with the accident initiator such as withdrawal of all control rods in case of UTOP accident and total loss of pumping power in the primary system for the case of ULOF accident. Then the calculation of total coolant flow-rate and flow distribution across the reactor core are performed which followed by the calculation of coolant and fuel temperature distribution, the calculation of energy and mass balance in the steam generator. Next, we solve the kinetic equation after calculating current feedback reactivities (Doppler, coolant density, core radial expansion and fuels axial expansion reactivity feedbacks) which then used in the calculation of amplitude and shape functions. The calculation is repeated by going back to the calculation of total coolant flow-rate and flow distribution across the core until reach the end of the simulation time.

3. Results and discussions

Table 1 shows main paramters for sample parameters for the current simulations. The reactor power is 800 MWt and this is a medium sized Modified CANDLE burn-up system.

Table 1 Main paramters used in this simulations

Parameter	Value
Power (MWth)	300
Core Geometry	2-D Cylinder
Refueling Periode(years)	10 years x10 batch
Fuel/cladding/coolant type	UN and PuN/SS316/Pb-208
Active core radius/height	117 cm/237.5cm
Reflector width (Pb-208)	70 cm
Fuel/structure/coolant volume fractions	60/12.5/27.5

The simulation results for Unprotected Loss of Flow (ULOF) accident case are shown in figures 1-4. Figure 1 shows coolant flow-rate in the core during unprotected loss of flow accident. Figure 2 shows power change with time during this accident. Fig. 3 shows The coolant decrease triggers imbalance between produced heat and coolant flow-rate causing temperature increase in coolant and fuel. Figure 3 shows coolant, cladding and fuel temperature change with time during the accident and finally Figure 4 shows reactivity feedback change with time during the accident. Temperature increase in coolant and fuel produces negative reactivity feedback which drive reactor power to decrease. After 100 seconds the coolant flow rate and the reactor power are approaching asymptotic levels and the total reactivity approaching zero. The maximum temperature in fuel and coolant are still far below safety limit.

The flow-rate decrease basically influenced by natural circulation level of the thermal hydraulic system of the NPP. About 100s after the accident begins, the core flow-rate has approached natural circulation levels and around 100 seconds the flow-rate change become relatively slow. In case of the reactor core power level, it also decreases following the decrease of the core flow-rate. However the asymptotic level is still at higher percentage compared to that of flow-rate.

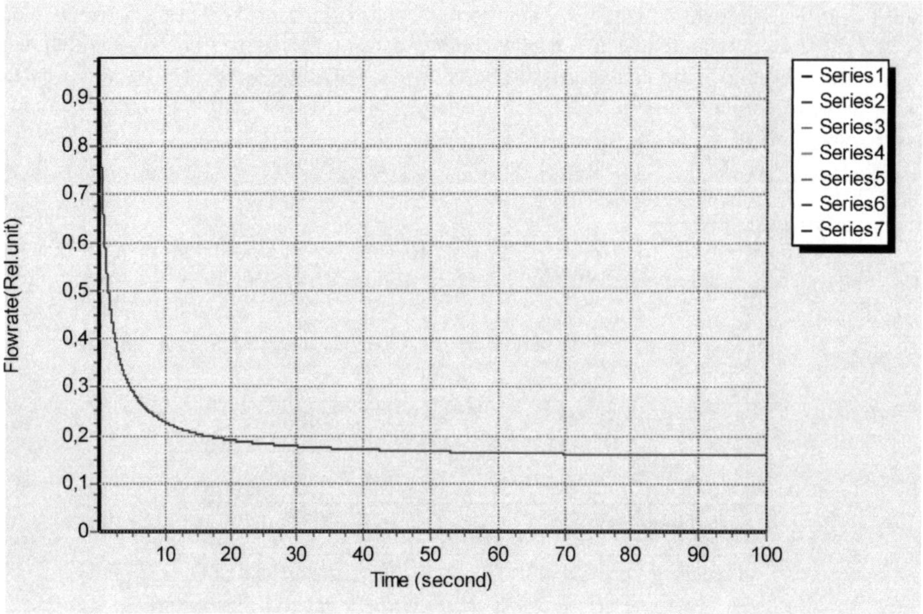

Figure 1 Coolant flow-rate change with time during loss of flow accident

After the decrease of the coolant flowrate, the fuel, cladding and cooolant temperatures initially increase and after reaching the peak then decrease and move toward new equilibrium level. The fuel temperature experience more significant reduction compared to the coolant and cladding temperature. This situation can be explained due to relatively large coolant-center fuel temperature different in the fuel pin. Due to the change of the fuel to coolant temperature difference during the accident the position of maximum fuel, cladding and temperatures are basically changing with the time and may not in the same positions for each coolant maximum temperature, cladding maximum temepratura, and fuel maximum temperature.

The reactivity feedbacks move in more complex situations. The Doppler and fuel axial expansion at the beginning give strong negative feedback, however, after reaching maximum temperature then their absolute values decerase and finally move toward positive direction. On the other hand, the core radial expansion and the coolant density reactivity feedback are definitely negative. However their values also decrease after reaching their maximum peak(absolute).

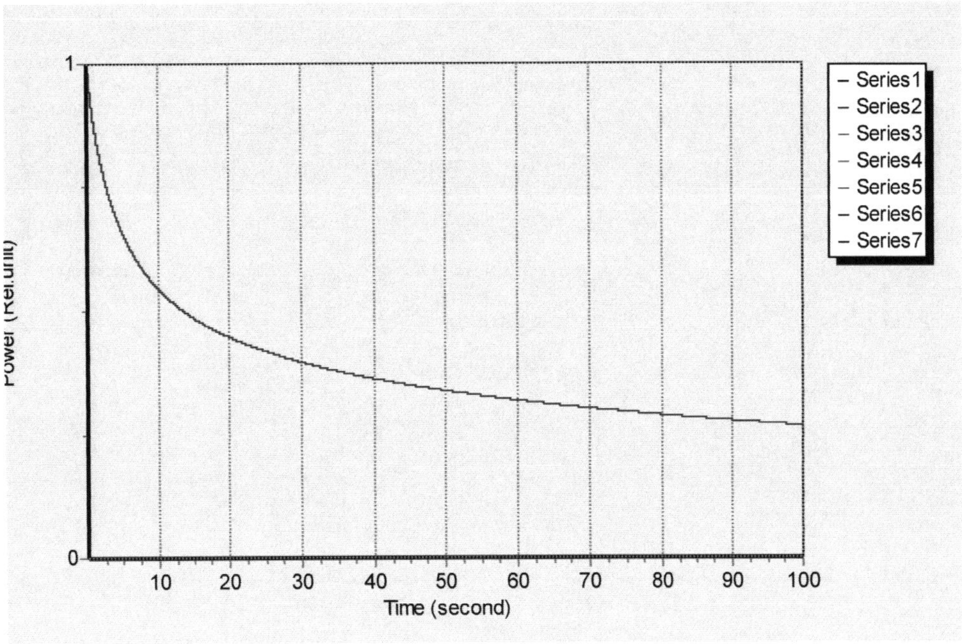

Figure 2 Power change with time during loss of flow accident

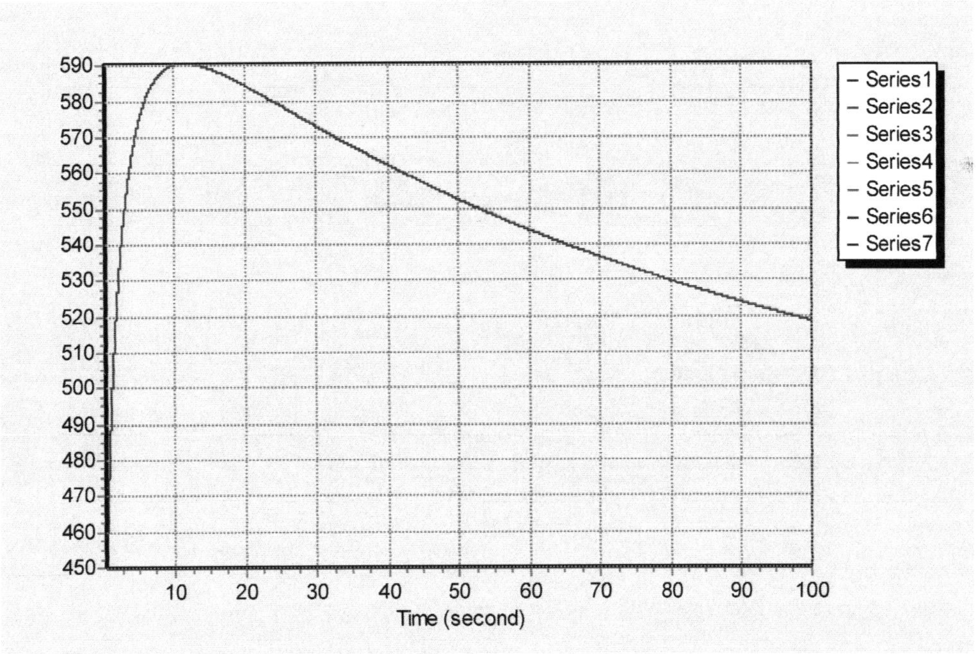

Figure 3 Hot spot temperatures change with time during loss of flow accident

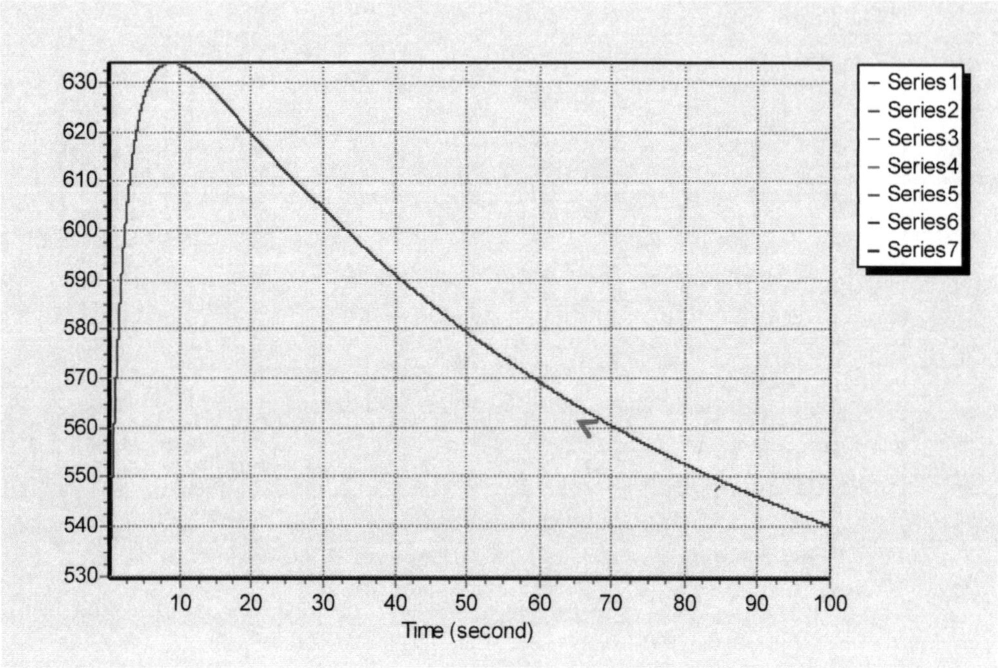

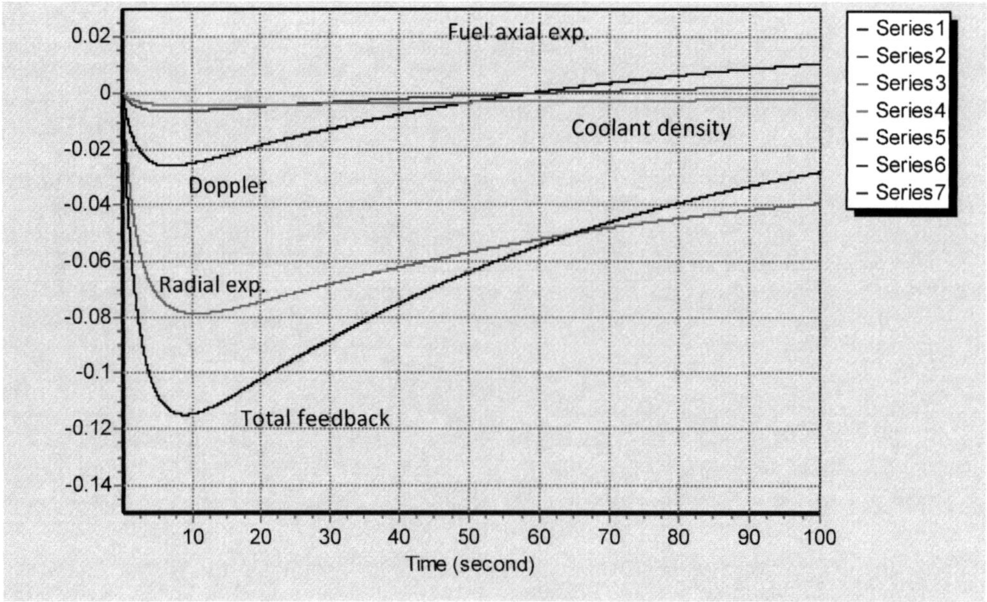

Figure 4 Reactivity feedback change with time during loss of flow accident

The accident analysis simulation results for Unprotected rod run-out Transient Over Power (UTOP) accident are shown in figures 5-8. As the accident intitator, the positive external reactivity triggers power increase which then causes temperature increase in coolant, cladding and fuel. Temperature increase in coolant, cladding and fuel produces negative reactivity feedback which is

IOP Conf. Series: Journal of Physics: Conf. Series **799** (2017) 012003 doi:10.1088/1742-6596/799/1/012003

used to gradually compensate external positive reactivity. After 100 seconds the power level approaching asymptotic levels and the total reactivity approaching zero. Doppler and Core radial expansion feedback play dominant role in UTOP accident. The temperature increase in fuel and coolant is still far below safety limit.

There are overshoot pattern of power following fast withdrawal of the control rod. Therefore this pattern influence the pattern of hotspot fuel, cladding and coolant temperature change with time and also the feedback. The most important difference between ULOF and UTOP cases are the fact that in the UTOP accident the fuel, cladding and coolant temperatures are continuously increase till approaching the asymptotic new power levels while in the ULOF case there are peak values of fuel coolant and cladding temperature as shown in Figures. 3 and 7. In the UTOP case the power is also increases till reaching asymptotic level while in the ULOF case decreases and then approaching asymptotic new power level as shown in Figs.3 and 6. In case of the UTOP accident the position change of maximum temperature for coolant, cladding and coolant are not significant.

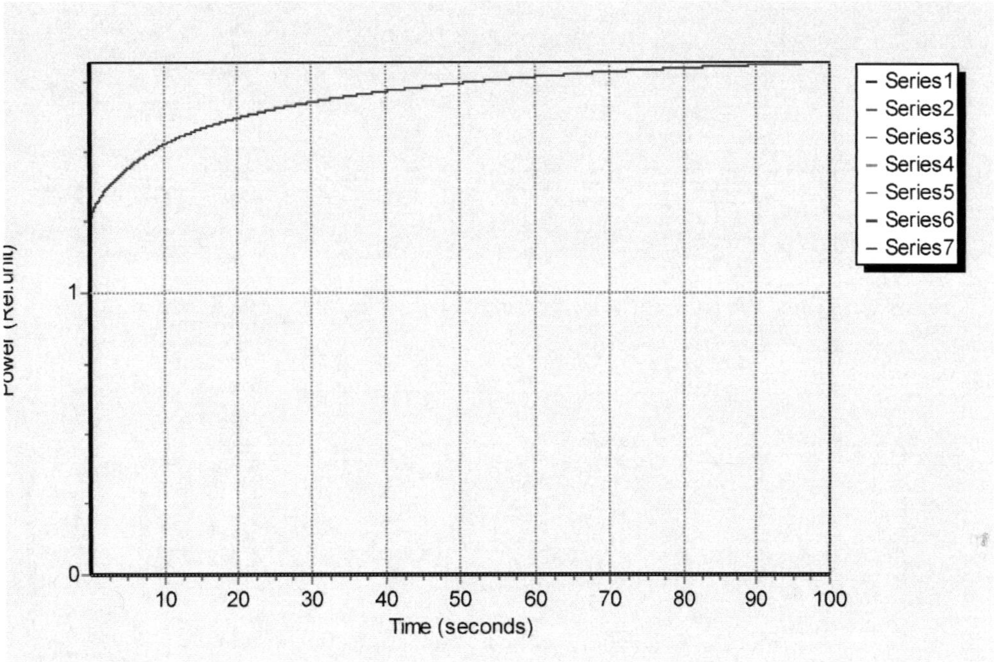

Figure 5 Power change with time during UTOP accident

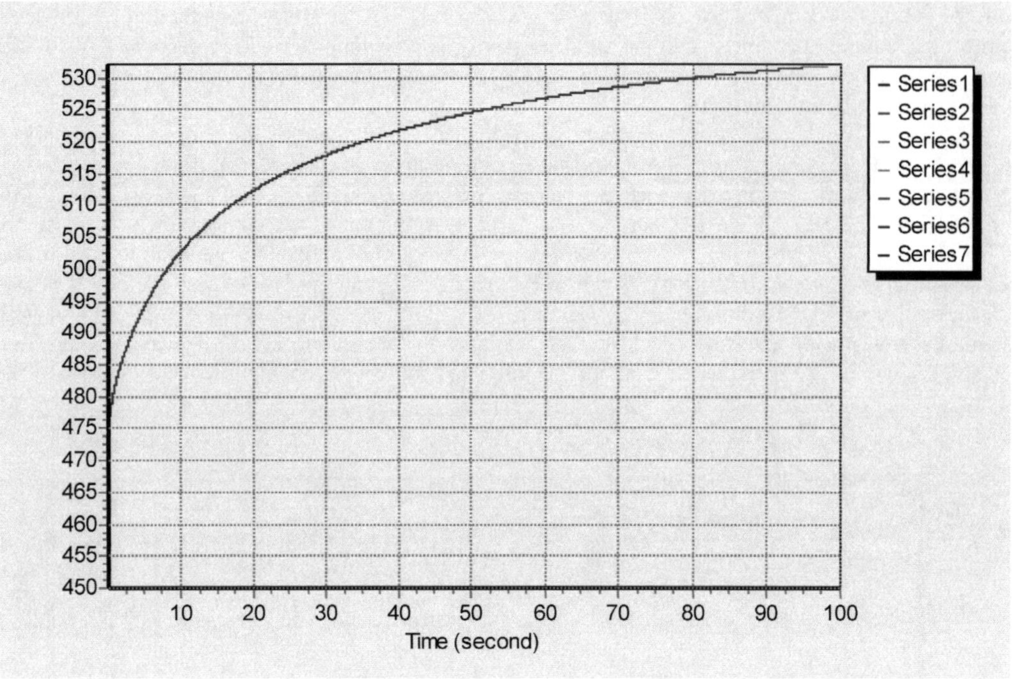

Figure 6 External and total reactivity feedback change with time during UTOP accident

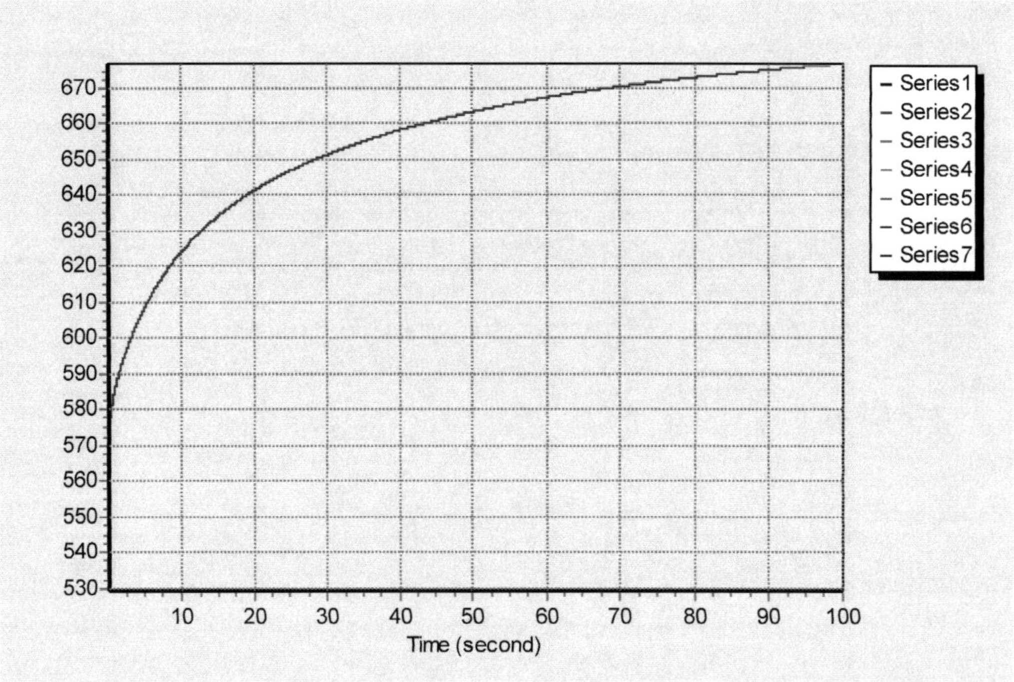

Figure 7 Hot spot temperatures chhange with time during UTOP accident

Figure 8 shows teh change of external and feedback reactivities during UTOP accident simulations. As shown in this figure, the new equilibrium conditions can be reached after total feedback reactivity can completely compensate external reactivity. From figure 8 it is shown that after 100 seconds from the accident begin the asymptotic level have been approached.

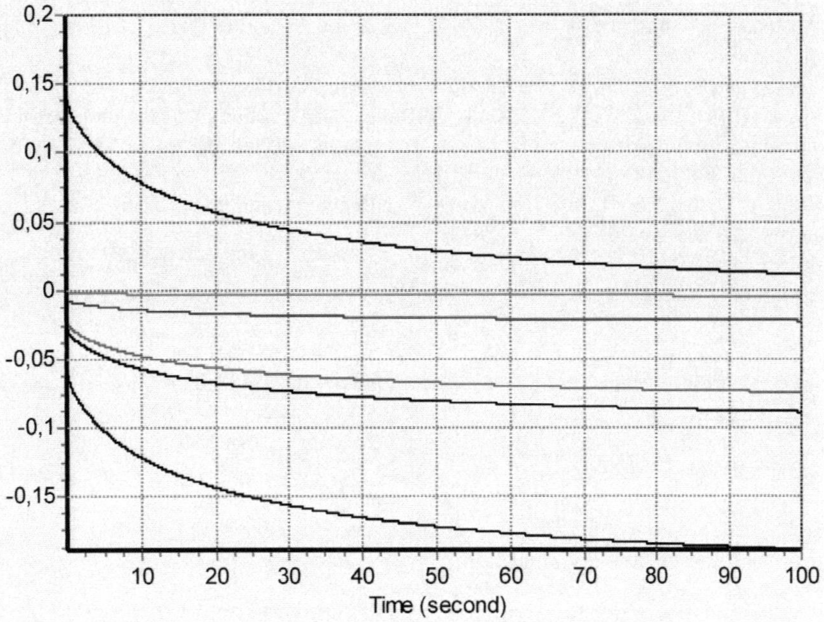

Figure 8 Reactivity change with time during UTOP accident

4. Conclusion

Safety analysis simulations have been performed for helium cooled 300MWt gas cooled fast reactors especially against unprotected loss of flow (ULOF) and unprotected rod-run out transient overpower (UTOP). Helium Cooled Small Modified CANDLE Reactors can survive ULOF and UTOP inherently. Natural circulation plays important role in the ULOF accident. Core radial expansion and Doppler reactivity feedback plays important role in the ULOF and UTOP accident.

References

[1] Zaki S,, 2008, Progress in Nuclear Energy, Neutronic Performance Comparation of MOX, Nitride and Metallic fuel based 25-100 MWe Pb-Bi Cooled Long Life Fast Reactors without on site Refuelling, Vol. 50, p. 276-278

[2] Zaki S, 2008, Progress in Nuclear Energy, Safety Performance Comparation of MOX, Nitride and Metallic fuel based 25-100 MWe Pb-Bi Cooled Long Life Fast Reactors without on site Refuelling, Vol. 50, p. 157-162

[3] Zaki Su'ud and H Sekimoto, 2010, IJNEST, Design study of long-life Pb-Bi cooled fast reactor with natural uranium as fuel cycle input using modified CANDLE burn-up scheme, Vol 5, No. 4 , p.347-368

[4] Su'ud, Z., and Sekimoto, H., 2012, IJNEST, Design study of medium-sized Pb-Bi cooled fast reactors with natural uranium as fuel cycle input using modified CANDLE burn-up scheme, Vol. 7 (1) , pp. 23-44.

[5] Zaki Su'ud and H Sekimoto, 2013, Annals of Nuclear Energy, The prospect of gas cooled fast reactors for long life reactors with natural uranium as fuel cycle input, Vol. 54 , pp. 58-66.
[6] S.K. Cheng and N.E. Todreas, 1986, Hydrodynamic models and correlationx for bare and wire-wrapped hexagonal rod bundles, Nucl. Eng. Des. 92, p 227.
[7] Zaki Su'ud and H. Sekimoto, 2010, IJNEST, Local Blockage analysis of Lead-Cooled Next Generation Nuclear Power Reactors, Vol. 5, No.2
[8] J.G. Guppy, 1993, Super System Code, NUREG/CR-3169.
[9] S. Kakac, R.K. Shah and W.Aung, 1987, Handbook of Single Phase Convective Heat Transfer, Wiley, pp.7.1-7.62.
[10] K.O. Ott and R.J. Neuhold, 1985, ANS, Nuclear Reactor Dynamics.
[11] H.Sekimoto and S. Zaki, 1995, Nucl. Technol. , Design study of lead and lead-bismuth cooled small long-life nuclear power reactors using metallic and nitride fuel, 109,p. 307-313.
[12] Zaki S. dan Sekimoto,H., 1996, Nucl. Eng. And Design, Accident Analysis of Lead or Lead-Bistmuth Cooled Small Safe Long-Life Fast Reactor using Metallic or Nitride Fuel, Vol. 162, Elsevier Science, pp.205-222.
[13] Zaki S., 1998, Progress of Nuclear Energy, Comparative Study on Safety Performance of Nitride Fueled Lead-Bismuth Cooled Fast Reactor with Various Power Level", Vol.32, No. ¾,, 571~577
[14] N. Nakagawa and K.Tsuchihashi, 1984, JAERI Memo, SLAROM: a code for cell homogenization calculation of fast reactor, (Japan Atomic Energy Research Institute, Ibaraki), JAERI M-1294

Delayed Neutrons Effect on Power Reactor with Variation of Fluid Fuel Velocity at MSR Fuji-12

Indarta Kuncoro Aji[1,*], Syeilendra Pramuditya[2], Novitrian[2], Dwi Irwanto[2] and Abdul Waris[2]

[1]Department of Physics, Faculty of Mathematics and Natural Sciences, Institut Teknologi Bandung,
Jl. Ganesha 10 Bandung 40132, INDONESIA
[2]Nuclear Physics & Biophysics Research Division, Department of Physics, Faculty of Mathematics and Natural Sciences,
Institut Teknologi Bandung,
Jl. Ganesha 10 Bandung 40132, INDONESIA

*Email: indartaaji@gmail.com

Abstract. As the nuclear reactor operate with liquid fuel, controlling velocity of the fuel flow on the Molten salt reactor very influence on the neutron kinetics in that reactor system. The effect of the pace fuel changes to the populations number of neutrons and power density on vertical direction (1 dimension) from the first until fifth year reactor operating had been analyzed on this research. This research had been conducted on MSR Fuji-12 with a two meters core high, and $LiF\text{-}BeF_2\text{-}ThF_4\text{-}^{233}UF_4$ as fuel composition respectively 71.78%-16%-11.86%-0.36%. Data of reactivity, neutron flux, and the macroscopic fission cross section obtained from ouput of SRAC (neutronic calculation code has been developed by JAEA, with JENDL-4.0 as data library on the SRAC calculation) was being used for the calculation process of this research. The calculation process of this research had been performed numerically by SOR (successive over relaxation) and finite difference methode, as well as using C programing language. From the calculation, regarding to the value of power density resulting from delayed neutrons, concluded that 20 m/s is the optimum fuel flow velocity in all the years reactor had operated. Where the increases number of power are inversely proportional with the fuel flow speed.

1. Introduction

Molten Salt Reactor is more unique than other Generation IV nuclear reactors concepts, because the fuel state of this reactor is liquid, and the reactor system based on the circulation of the molten salt in the core reactor and the heat exchanger. In concequence, the neutron kinetics calculation analysis of the MSR is difference with the other type of reactor.

The neutron kinetics calculation study of MSR has been discussed in the literature [1-3]. On these publications have described that the neutron kinetics calculation procedure of MSR is different from that of the conventional reactors are use solid fuel.

The number of delayed neutron precursors influenced by the fuel velocity. However, under the steady state condition, the fuel salt velocity has little effect on the fast and the thermal fluxes [2]. Study on effect during transient condition, respectively; the pump coastdown transient, rods drop transient, and the inlet temperature drop transient against the relative power had been calculated by the code DRAGON [2]. Delayed neutron and temperature feedback is useful in providing an estimate of the transient behavior of the reactor power [3,4].

Many scientist have focused in short time scale of the neutron kinetics study, to analyze the safety system reactor, such as disturbance or accident; reactor start up; and short modeling experiment [2-5]. However, study about neutron kinetics in long time scale is also important on this research area, because it has related with nuclear fuel cycle study; fuel burnup and waste management. The ability to reduce nuclear waste and nuclear weapon materials is one advantages by MSR [6, 7]. This paper describes the solution of the neutron kinetics study, especially delayed neutron and power reactor have produced when the reactor operate during five years.

2. MSR FUJI-12 Descriptiom

The main features of MSR Fuji-12 is shown at Table 1. Fuji-12 is the smallest MSR has been developed for energy production with total power output around 100 MWe [8]. The core size is shown in the third column of Table 1. Fuji-12 core consists from hexagonal cylinders assembly with flow channel for fuel in the middle, and it has made from graphite with number density is 1.84 g/cm^3 [9]. Equivalent diameter of the assembly is 20 cm, and the flow channel diameter is 10.95 cm. Graphite assembly is also functioned as moderator with several conditions; low absorption cross sections, resistant to height temperatures, high thermal conductivity, and has fine mechanical strength.

Figure 1 interpretes that the lifetime of this sistem is more than 15 years, it is depended by the strength of the graphite. Refueling times is influenced by the reactivity of the reactor, where the fissile materials composition has important role to the number of reactivity [10].

Tabel 1. Specifications of Fuji-12 [1, 10]

Design Parameters	Specifications
Thermal output	350 MW
Efficiency	40%
Core Geometry	
High	400 cm
Diameter	400 cm
Fuel Salt Composition	
LiF	71.78 mol %
BeF_2	16.00
ThF_4	11.86
$^{233}UF_4$	0.36
Fuel Salt Temperature	
Inlet	833 K
Outlet	973 K
Refueling Time	5 years
Average Power Density	7 kWt/ liter

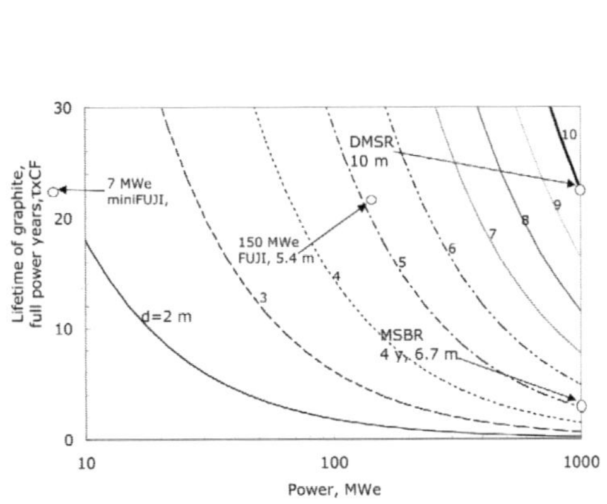

Figure. 1. Lifetime of graphite core is related to core size and power of MSR [11, 12]

3. Theory and Calculation

On the molten salt reactor, delayed neutron precursors drift with the fuel salt flow. This conditions make the neutron kinetics characteristics of the molten salt reactor difference than others solid-fuel reactors. The neutron kinetics equations with one group of flux neutron are given;

$$\frac{\partial n}{\partial t} + \nabla \cdot (Un) = \beta v \Sigma_f \phi(r,z,t) + \Sigma_{i=1}^{6} \lambda_i C_i(r,z,t) \qquad (1)$$

$$\frac{\partial C_i}{\partial t} + \nabla \cdot (UC_i) = \beta_i v \Sigma_f \phi(r,z,t) - \lambda_i C_i(r,z,t) \qquad (2)$$

On the steady state conditions, time has no affect on the neutron density and the average density of delayed neutron precursors, so that we can assume as; $\frac{\partial n}{\partial t} = 0$ and $\frac{\partial C_i}{\partial t} = 0$. Moreover, those equations can be simplified because the calculation method of this research is only considering fuel displacement at vertical direction (z axial). Where the fuel salt is moving form bottom to top of chanels at the core reactor. Respectivelly, the equation are given;

$$U_z \frac{\partial n}{\partial z} = \beta v \Sigma_f \phi(z) + \Sigma_{i=1}^{6} \lambda_i C_i(z) \qquad (3)$$

$$U_z \frac{\partial C_i}{\partial z} = \beta_i v \Sigma_f \phi(z) - \lambda_i C_i(z) \qquad (4)$$

Boundary condition a given from bottom to top by the core reactor. When the fuel salt is circulating with velocity (U_z), circulate time of fuel salt in the external loop can be defined as (τ_L) [5,9]. Reduction the average density of delayed neutron precursors at the fuel salt in the external loop is depended by decay constant and external loop transit time, mathematically can be difined at Eq. (5). For the neutron flux, boundary condition can be assumed as ($\phi(z) = 0$) at the bottom and top of the core reactor.

$$\frac{dC(\tau_L)}{d\tau} = \lambda C(\tau_L) \qquad (5)$$

When the high of active core is 2a, accordingly the boundary condition for neutron flux and density of delayed neutron precursors read as;

$$\phi(-a) = \phi(a) = 0 \qquad (6)$$

Tabel 2. Delayed neutron fractions and the precursors decay constants [15]

Group	β_i	λ_i
1	2.58	0.0129
2	6.96	0.0347
3	5.46	0.119
4	10.9	0.288
5	3.59	0.805
6	1.26	2.47

$$C(-a) = C(a)e^{-\lambda\tau_L} \quad (7)$$

When the reactor operate on the thermal neutron with U^{233} as fuel, the power produced from delayed neutron can be calculated from the following equation [14];

$$P(W/cm^3) = \frac{\Sigma_f\, w\, n}{3.17x\, 10^{10}} \quad (8)$$

4. Results and Discussion

As same as with prompt neutron, fission reaction is also occure on delay neutron. Analysis of the delayed neutron in the core reactor is assisting in investigating the characteristic of the core reactor. Changes velocity of the fuel flow affected on the power had been generated by delayed neutron. Power dropped exponentially when the velocity of fuel flow was being increased. Shown in figure 2. The reactor had operated in five years, and generally all graphics of the power for every years dropped exponentially. One of the interesting points of the graphics is when the reactor operate at 0.2 m/s of fuel flow velocity. The power significantly decreas at 0.05 m/s to 0.2 m/s of fuel flow velocity, and slopingly at 0.2 m/s to 0.5 m/s.

Basically, changes of the power was affected by the number of neutrons at the core reactor. Where the number of neutrons was affected by fluid fuel velocity.

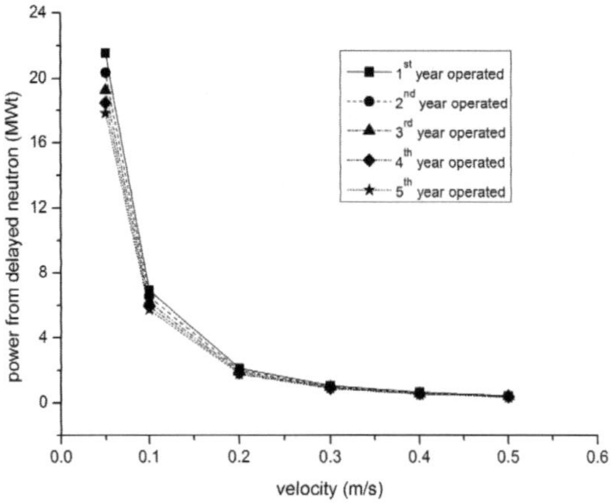

Figure. 2. Power was produced by delayed neutron with several various speed of fuel flow.

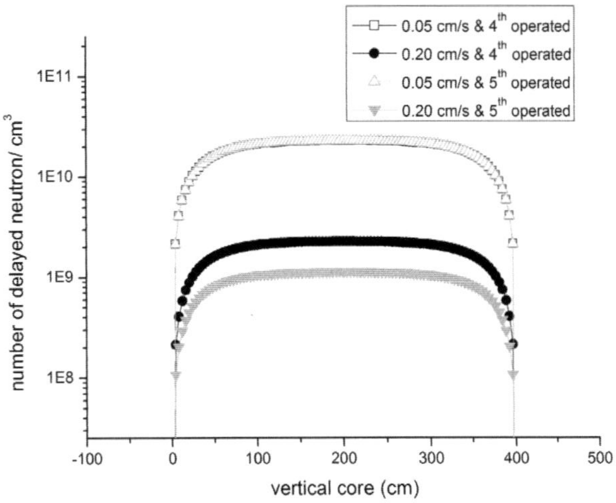

Figure. 3. Distribution of number of delayed neutrons in the vertical core on the fourth and fifth years reactor had operated.

Distribution of the neutrons is shown at figure 3. The distribution number of delayed neutron between fourth and fifth years operation when the reactor had operated on 0.05 m/s fuel flow velocity are similar. But, it is look difference for distribution number of delayed neutron has operated on 0.2 m/s.

On the SRAC calculation was found that the critically condition occur on the fourth year reactor operated with number multiplication effective is 1.01547, and the number multiplication effective on the fifth year reactor operated is 0.98527. It because reduction number of fissile materials impacting on reduction number of neutrons (prompt and delayed). So that, when the reactor operated on the subcritical condition (on the fifth year operation), it status can be increased on critical condition with reducing velocity of the fuel flow, and assuming the availability of fissile materias are still quite.

On the other hand, can be conclude that increasing the reactor status from subcritical to critical condition on the fifth year reactor operated will increasing the utilization of fertile materials (more efficient). So that, the number of fertile materials on the spent fuel will decrease.

5. Conclusion

The result of this research show that velocity of the fuel salt has effect on number of delayed neutron distribution at core reactor and the power had been produced. Optimization the utilization of fuel materials (fissile materials) is a benefit from reactor with fluid fuels. From the result can be predicted that the 0.2 m/s is the normal condition by velocity of fluid fuel during reactor has operated.

References

[1] T. Yamamoto, K. Mitachi, and T. Suzuki, Steady Sate Analysis of Small Molten Salt Reactor (Effect of Fuel Salt Flow on Reactor Characteristics, *JSME Int. Journal,* **48**, 610, 2005.

[2] D. L. Zhang *et al.*, Analysis on The Neutron Kinetics for A Molten Salt Reactor, *Progress in Nuc. Energy,* **51**, 624, 2009.

[3] A. Cammi *et al.*, Dimensional Effect in The Modelling of MSR Dynamics; Moving on From Simplified Schemed of Analysis to A Multi-physics Modelling Approach, *Nuc. Eng. And Design,* **246,** 12, 2012.

[4] A. A. Nahla, An Analytical Solution for The Point Reactor Kinetics Equations With One Group of Delayed Neutrons and The Adiabatic Feedback Model, *Progress in Nuc. Energy,* **51,** 124, 2009.

[5] I. Pazsit, V. Dykin, and R. Sanchez, The Point Kinetic Component of Neutron Noise in An MSR, *Annals of Nuc. Energy,* **64**, 344, 2014.
[6] A. Waris, *et al.*, Preliminary Study on Plutonium and Minor Actinides Utilization in Thorium-nes Minifuji Reactor, *Energy. Conv. and Management,* **72**, 27, 2013.
[7] I. K. Aji, and A. Waris, Int. Conf. on Advances in Nuclear Science and Engineering (Sep. 16-19, 2013, Denpasar, Indonesia), vol 1615 (New York: AIP Conf. Proc), 57, 2014.
[8] O. Benes, *et al.*, Assessment of Liquid Salts for Innovative Applications, D50-V4, (Europian Commission Under The Euratom Research and Training Program), 2009.
[9] T. Yamamoto, *et al.*, Transient Characteristics of Small Molten Salt Reactor During Blockage Accident, *Heat Transfer-Asian Research,* **35**, 434, 2006.
[10] A. Waris, *et al.*, Comparative Study on ^{233}U and Plutonium Utilization in Molten Salt Reactor, Indonesian Journal of Physics, **21**, 77, 2010.
[11] K. Furukawa, *et al.*, A Road Map for The Realization of Global Scale Thorium Breeding Fuel Cycle by Single Molten Fluoride Flow, Molten Salt Reactor for Sustainable Nuclear Power-MSR FUJI, *Energy. Conv. and Management,* **49**, 1832, 2008.
[12] R. W. Moir, Reccomendation for A Restart of Molten Salt Development, *Energy. Conv. and Management,* **49**, 1849, 2008.
[13] A. Cammi *et al.*, A Multi-Physics Modelling Approach to The Dynamics of Molten Salt Reactors, *Annals of Nuc. Energy,* **38**, 1356, 2011.
[14] A. E. Waltar *et al.*, Fast Breeder Reactor, Pergamon Press, New York, 187, 1976.
[15] K. Okumura *et al.*, SRAC (Ver. 2002); The Comperhensive Neutronics Calculation Code System - Volume I, Japan Atomic Energy Research Institute (JAERI), 2002.
[16] J. J. Duderstad, and L. J. Hamilton, Nuclear Reactor Analysis, John Wiley & Sons Inc., 1976.
[17] W. M. Stacey, Nuclear Reactor Physics, John Wiley & Sons Inc., 2001.

IOP Conf. Series: Journal of Physics: Conf. Series **799** (2017) 012005 doi:10.1088/1742-6596/799/1/012005

Impact of New Nuclear Data Libraries on Small Sized Long Life CANDLE HTGR Design Parameters

Peng Hong Liem[1], **Donny Hartanto**[2], **Hoai Nam Tran**[3]

[1]Nippon Advanced Information Service (NAIS Co., Inc.) Scientific Computational Division, 416 Muramatsu, Tokaimura, Ibaraki, Japan
[2]Fast Reactor Design Division, Korea Atomic Energy Research Institute (KAERI), 111, Daedeok-daero 989, Yuseong-gu, Daejeon 34057, Korea
[3]Institute of Research and Development, Duy Tan University, K7/25 Quang Truang, Da Nang, Vietnam

Email: liemph@nais.ne.jp

Abstract. The impact of new evaluated nuclear data libraries (JENDL-4.0, ENDF/B-VII.0 and JEFF-3.1) on the core characteristics of small-sized long-life CANDLE High Temperature Gas-Cooled Reactors (HTGRs) with uranium and thorium fuel cycles was investigated. The most important parameters of the CANDLE core characteristics investigated here covered (1) infinite multiplication factor of the fresh fuel containing burnable poison, (2) the effective multiplication factor of the equilibrium core, (3) the moving velocity of the burning region, (4) the attained discharge burnup, and (5) the maximum power density. The reference case was taken from the current JENDL-3.3 results. For the uranium fuel cycle, the impact of the new libraries was small, while significant impact was found for thorium fuel cycle. The findings indicated the needs of more accurate nuclear data libraries for nuclides involved in thorium fuel cycle in the future.

1. Introduction

Recently pursued innovative reactor design such as CANDLE (Constant Axial shape of Neutron flux, nuclide densities and power shape During Life of Energy producing reactors) [1] requires an assessment of the accuracy of their main reactor design parameters to determine properly the design feasibility, their safety margins, and in the future their economical margins. Innovative reactor designs often adopt a novel way of fuel burning scheme and fuel compositions which result in very heterogeneous core and/or blanket regions involving complex spatial power as well as different neutron spectra distributions across the reactor. In addition, the designs are commonly aimed at a very high discharged fuel burnup which require accurate reactor physics constants for nuclear transmutation whose accuracy is naturally rooted back to the high quality of the evaluated nuclear data used.

Several new evaluated nuclear data libraries have been released, namely the JENDL-4.0 [2] (replacing the older version of JENDL-3.3), ENDF/B-VII.0 [3] and JEFF-3.1 [4] etc, which cover more number of nuclides and have better agreement with integral measurement results. The release of the new evaluated nuclear data was followed by compilation works of code-specific working libraries for productive reactor design and analysis. For example, after Japan Atomic Energy Agency (JAEA) released the JENDL-4.0, the Agency also prepared, tested, verified and validated the JENDL-4.0 based SRAC2006 library [5]. SRAC2006 [5] used in the present work is a comprehensive neutronics calculation code system which consists of many modules from transport based cell calculation module to multidimensional diffusion based whole core neutronics calculation module.

In the present paper, the impacts of new evaluated nuclear data libraries (JENDL, ENDF/B and JEFF) on the core characteristics of the innovative CANDLE reactor design are investigated. Although the novel CANDLE burnup scheme can be applied both for fast and thermal reactors, in the present work, we consider prismatic (block) type high temperature gas-cooled reactors (HTGR) which operate under thermal neutron spectra. Two fuel cycles are considered for the CANDLE HTGRs, namely the uranium and thorium fuel cycles. Thorium fuel cycle are not commonly used presently, however in the future, the thorium potential would be utilized and investigation on the impact of the new evaluated nuclear data libraries on the thorium fuel cycle performance is of great interest. Both the CANDLE HTGRs with uranium and thorium fuel cycles may reach very high discharge burnup levels of more than 100 GWd/t.

2. CANDLE high temperature gas-cooled reactor

2.1. CANDLE Concept

The innovative CANDLE (Constant Axial shape of Neutron flux, nuclide densities and power shape During Life of Energy producing reactor) burning scheme was proposed by Sekimoto et al. [1] originally aimed for fast reactors. Under this burning scheme, the burning region (sometime it is called burning wave) moves autonomously with a constant velocity along the core axis from bottom to top (or from top to bottom) as shown in Fig. 1. As shown in the figure, the core can be roughly divided into three regions: (1) fresh fuel region (kinf <1), (2) burning region (kinf >1) and (3) spent fuel region (kinf >1). When the burning scheme is applied to a prismatic/block-type HTGR, burnable poison (for e.g. natural gadolinium) is used to adjust the kinf of the fresh fuel to be sub-critical.

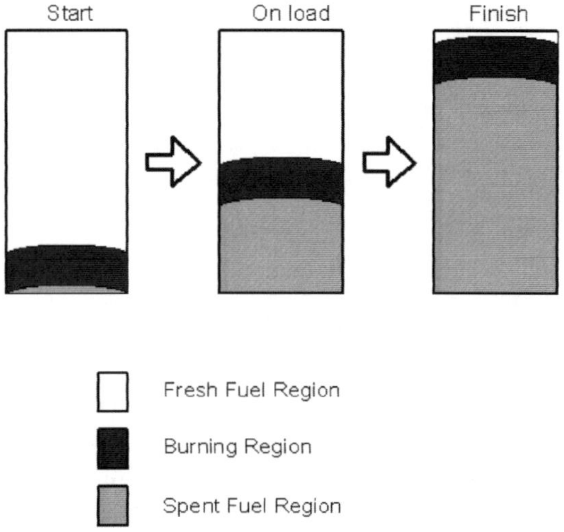

Figure 1. CANDLE burnup concept.

The CANDLE HTGR burn-up process is as follows (Fig. 2). Neutrons leaked from the burning region into the fresh fuel region will be absorbed by the burnable poison and the burning region will move slowly into the fresh fuel region with depleted burnable poison. In the burning region, depletion of fissile material for energy production is accompanied by conversion of fertile material into fissile material. The spent fuel region is the region left by the burning region which contains mainly fission products and depleted fuel. After the CANDLE HTGR is operated for a certain core life time, the reactor can be shut down for refueling. If the core active height is design properly the CANDLE concept may feature a long life HTGR design as will be shown later.

For a unique combination of core geometry and fresh fuel composition, one can find an equilibrium critical condition where CANDLE burning scheme is realized. Under the equilibrium condition, the moving (axial) velocity of the burning region is constant. Analytical codes for obtaining either the equilibrium condition or for simulating the reactor start-up (running-in phase) have been developed. In our previous works as well as in the present work we consider small sized long life prismatic/ block-type HTGRs adopting CANDLE burning scheme which can take full advantages of CANDLE properties:

1. Constant reactor parameters (e.g. power peaking, reactivity coefficients etc.) during reactor operation. This will simplify not only the reactor design itself and its licensing process but also simplify its reactor operation and maintenance.
2. No requirement for burn-up reactivity control mechanism. Besides simplifying the reactor design, a severe control rod ejection accident during full power operation (under nominal pressure) can be avoided.
3. Proportionality of core height to reactor core life. A long life core can be easily designed by adjusting the core height.
4. Sub-criticality of fresh fuel. No criticality accident will occur during transportation and storage of fresh fuels.

In addition, application of CANDLE burning scheme to small sized long life HTGRs can be realized by the present coated fuel particle and HTGR reactor technologies. As for the fuel cycle, CANDLE HTGRs can be applied for uranium [6] and thorium [7] fuel cycles. In our previous work [7], we showed that CANDLE HTGRs with thorium fuel gave better burnup performance than the ones with uranium fuel. This fact is also true for other thorium fueled HTGRs of the pebble bed type with once-through-then-out (OTTO), multipass as well as peu-a-peu (PAP) schema as described, analyzed and reviewed comprehensively by Liem et al. [8].

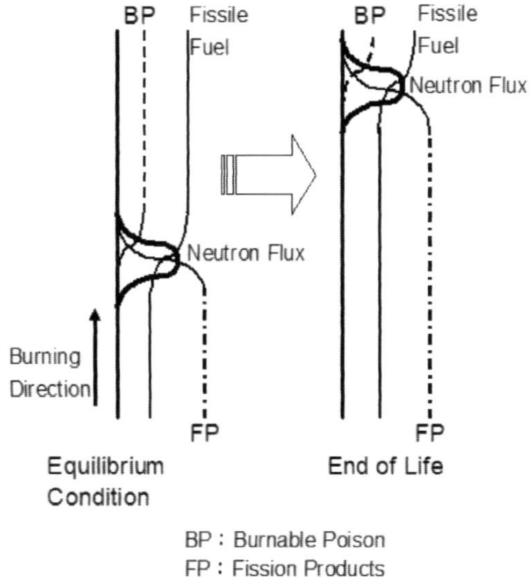

Figure 2. CANDLE burnup application to HTGR.

2.2. CANDLE Analytical Code and Group Constants

An in-house analytical tool for obtaining the CANDLE burnup equilibrium condition has long been developed in our previous works for both fast and thermal reactors. The code takes into account the burning region movement, nuclides burnup and criticality equations simultaneously. The details of the computational procedures are given in the references by Ohoka and Sekimoto [6] and Ismail et al. [7]. In this section the group-constants preparation, CANDLE burnup equilibrium and critical search procedure as well as the calculation conditions are discussed since they are directly affected the by the evaluated nuclear data library used.

The analytical tool used for obtaining the CANDLE burnup equilibrium condition, in principal, needs (1) effective microscopic cross sections and (2) burnup related data such as fission product yields, decay constants, branching ratios and a depletion chain. The effective microscopic cross sections are prepared using the collision probability (PIJ) module of SRAC2006 code system with a SRAC library based on a particular evaluated nuclear data. For the present version of SRAC2006, the available SRAC libraries are based on JENDL-3.3 [11], JENDL-4.0, ENDF/B-VII.0 and JEFF-3.1 evaluated nuclear data libraries. The burnup related data are taken directly from the SRAC library.

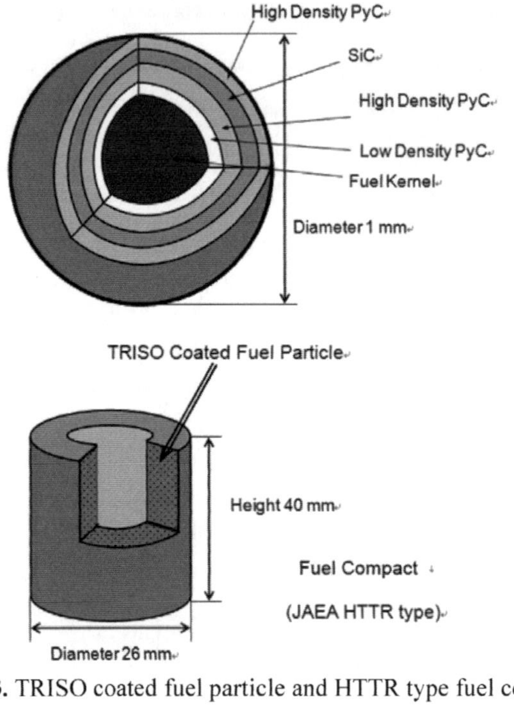

Figure 3. TRISO coated fuel particle and HTTR type fuel compact.

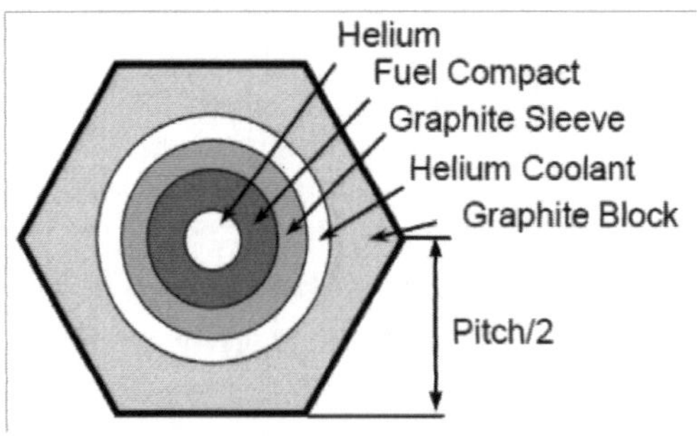

Figure 4. Cell calculation model by SRAC2006 code system.

Figs. 3 and 4 show the SRAC2006 calculation model for the group-constants generation. Since we adopted the JAEA High Temperature Engineering Test Reactor (HTTR, 30 MWth) [9,10] type HTGR fuel, in the 2-D hexagonal fuel lattice, the cell is divided into annulus fuel compact, graphite sleeve, annulus helium coolant channel and graphite block. Use of TRISO coated fuel particles in the HTGR fuel compact demands double heterogeneity calculation feature which is provided by SRAC code system in its PIJ module. Furthermore, in the resonance energy region, the ultra-fine energy group capability of SRAC (PEACO module with its MCROSS ultrafine group library) was utilized to obtain accurate effective accurate cross sections on the energy region.

Table 1. Neutron energy group structure for CANDLE calculations (4-group).

Group		Energies (eV)	
		Upper	Lower
1	Fission	1.0000E+07	1.1109E+05
2	Slowdown	1.1109E+05	2.9023E+01
3	Resonance	2.9023E+01	2.3824E+00
4	Thermal	2.3824E+00	1.0000E-05

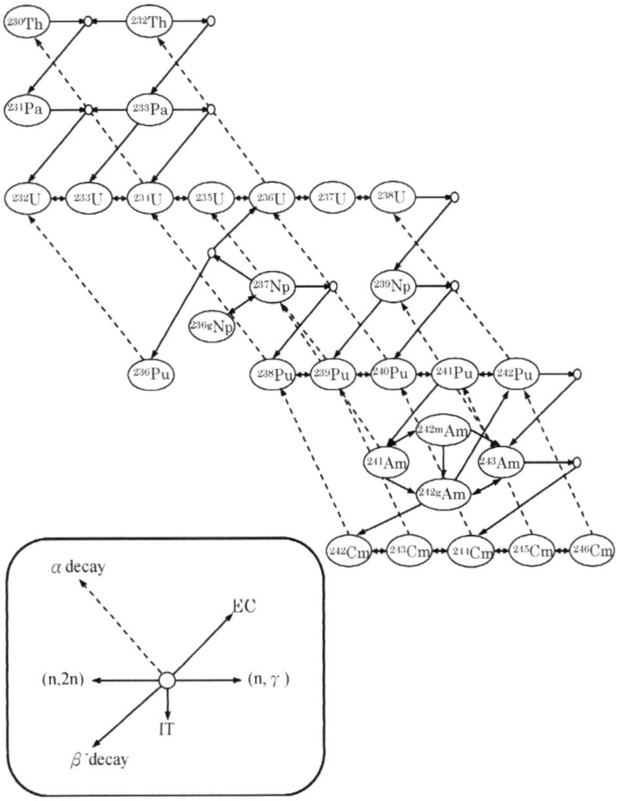

Figure 5. Burnup chain used for CANDLE HTGR burnup calculations (heavy metals).

After obtaining the effective microscopic cross sections in 107 energy group (the largest number for SRAC code and library) then the cross sections are collapsed into 4 energy group (Table 1) to be used for the CANDLE burnup calculations. The depletion chain which consists of 29 heavy metal, 66 important fission product (including one pseudo fission product) nuclides and 16 burnable poison nuclides is based on SRAC's THCM66FP chain [5] shown in Fig. 5.

2.3. CANDLE Burnup Equilibrium and Critical Search Procedure

In order to obtain a CANDLE burnup equilibrium and critical condition, in the first stage, one has to determine the fresh fuel kernel composition, i.e. the fissile U-235 or U-233 enrichment, natural Gd burnable poison, and search the CANDLE burnup equilibrium condition.

In this stage, the obtained equilibrium condition may give an effective neutron multiplication factor (keff) which is not critical. In the second stage, only the natural gadolinium burnable poison concentration is adjusted to obtain a critical (keff=1.0) and equilibrium condition of CANDLE burnup. This two-stage iterative procedure is terminated if the keff is near 1.0 within 1 % convergence criterion. There is a possibility that a certain composition of U-235 or U-233 enrichment and Gd burnable poison weight fraction will not give a critical and equilibrium condition of CANDLE burnup.

Table 2. Design parameters of small sized long life thorium CANDLE high temperature gas-cooled reactor.

Thermal power (MWth)	30
Core	
Diameter (cm)	230
Active height (cm)	800
Radial reflector (graphite)	
Thickness (cm)	100
Coated fuel particle	
Fuel	(U-235/U-238)O_2 or (U-233/Th-232)O_2
Uranium enrichment (w/o)	< 20%
Burnable poison material	Natural gadolinium
Type	TRISO
Kernel diameter (mm)	0.608
Particle diameter (mm)	0.940
Coating material	PyC / PyC / SiC / PyC
Thickness (mm)	0.060 / 0.030 / 0.030 / 0.046
Density (g/cm^3)	1.143 / 1.878 / 3.201 / 1.869
Packing fraction (v/o)	30.0
Fuel compact	JAEA HTTR type
Inner diameter (cm)	1.00
Outer diameter (cm)	2.60
Graphite sleeve	
Inner diameter (cm)	2.60
Outer diameter (cm)	3.40
Coolant annulus channel	
Inner diameter (cm)	3.40
Outer diameter (cm)	4.10
Fuel pitch	
Flat to flat distance (cm)	6.60

2.4. Main Design Parameters and Calculation Conditions

In order to investigate the impact of the new evaluated nuclear data libraries on the core characteristics of the CANDLE HTGRs we selected our previous small sized, long-life CANDLE HTGR design as a reference where the design adopted both uranium and thorium fuel cycles. The reactor main design parameters are shown in Table 2. This reference design, up to now, was calculated by using JENDL-3.3 based SRAC library. To investigate the impact of the new libraries, we selected the available SRAC libraries, namely the JENDL-4.0, ENDF/B-VII.0 and JEFF-3.1 based ones.

The most important parameters of critical CANDLE core characteristics are (1) infinite multiplication factor (kinf) of the fresh fuel containing burnable poison, (2) the effective multiplication factor (keff) of the equilibrium core, (3) the moving velocity of the burning region (Vel), (4) the attained discharge burnup (BU), and (5) the maximum power density (Qmax). In order to clearly observe the impact of the new libraries on these parameters, we adopted the fresh fuel composition of the reference case (calculated with JENDL-3.3 library) and used the fresh fuel composition for other libraries and keeping other calculation conditions unaltered.

3. Impact of new evaluated nuclear data libraries

Firstly, the core characteristics of the CANDLE HTGR for the reference case i.e. the ones calculated by using the JENDL-3.3 library is shown in Table 3. For both uranium and thorium fuel cycles, four values of fissile (U-235 or U-233) enrichment were evaluated. The smallest enrichment (6.5 w/o) was determined from the criticality requirement of the design, while the highest enrichment (19.75 w/o) was attributed to the non-proliferation issue that is the use of LEU. In the table, for a particular fissile enrichment the unique value of burnable poison (natural Gd) concentration needed to establish an equilibrium and critical CANDLE HTGR is also shown. Since a higher fissile enrichment will result in a higher initial kinf (without Gd) then the required Gd concentration increases with higher fissile enrichment. Compared to uranium fuel cycle, the same fissile enrichment of U-233 provides more reactive composition than the U-235 does so that higher fuel burnup can be expected.

Calculation results of representative example of CANDLE HTGR with uranium fuel cycle and thorium fuel cycle are shown in Figs. 6 and 7, respectively (fissile enrichment of 15 w/o). In Fig. 6, the burning region (wave) is moving from right to left approximately 30 cm/year, and the depleted fuel in the right side of the figure has the average burnup of approximately 100 GWd/t. The maximum power density is less than 5 W/cm3 which assures the safety of the reactor during normal and accident conditions.

Table 3. Calculated main parameters of CANDLE HTGR core characteristics with uranium and thorium fuel cycles (reference case, JENDL-3.3).

Fuel Cycle	Enrichment (w/o)	Gd (w/o)	kinf	keff	BU (GWd/t)	Vel. (cm/year)	Qmax (W/cm^3)
Uranium	19.75	3.80	0.66223	1.00970	121.8	24.2	4.75
	15	2.90	0.58728	1.00146	98.6	29.7	4.75
	10	1.70	0.50888	1.00394	65.5	44.1	4.61
	6.5	0.90	0.45136	1.00253	42.5	67.4	4.09
Thorium	19.75	11.70	0.87426	1.00187	175.1	18.5	5.37
	15	8.75	0.79043	1.00164	138.4	22.7	5.93
	10	5.40	0.66629	1.00300	95.5	31.8	6.10
	6.5	2.90	0.54432	1.00249	57.7	51.1	6.04

As discussed above, the thorium fuel cycle (cf. Fig. 7) shows better core characteristics in term of achievable fuel burnup and much slower burning region (wave) moving velocity. Slower velocity allows the reactor designer to produce longer core life time for the same effective core height. The downside of thorium fuel cycle compared with the uranium one is a higher maximum power density. Although in the present small long life CANDLE HTGR design, the absolute values are low, the maximum power density would be problematic for large power CANDLE HTGRs.

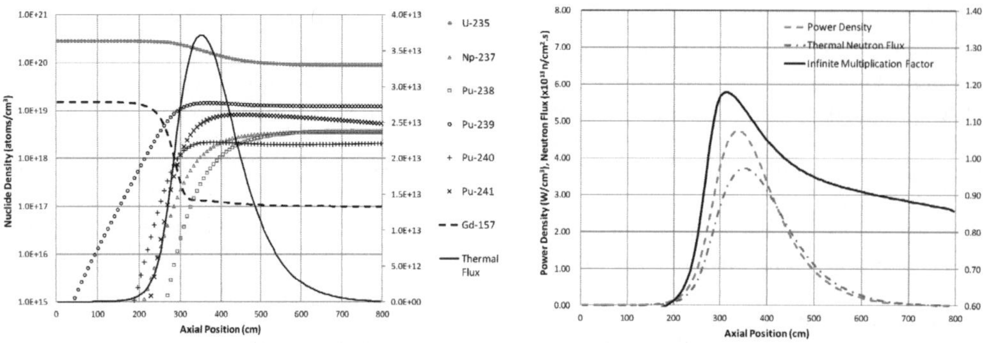

[Left] Nuclide densities (left axis) and thermal group neutron flux (right axis)
[Right] Power density, thermal group neutron flux (left axis) and kinf (right axis)
Figure 6. Critical and equilibrium condition of CANDLE HTGR with U-235 enrichment of 15 w/o and burnable poison Gd concentration of 2.90 w/o (reference case, JENDL-3.3).

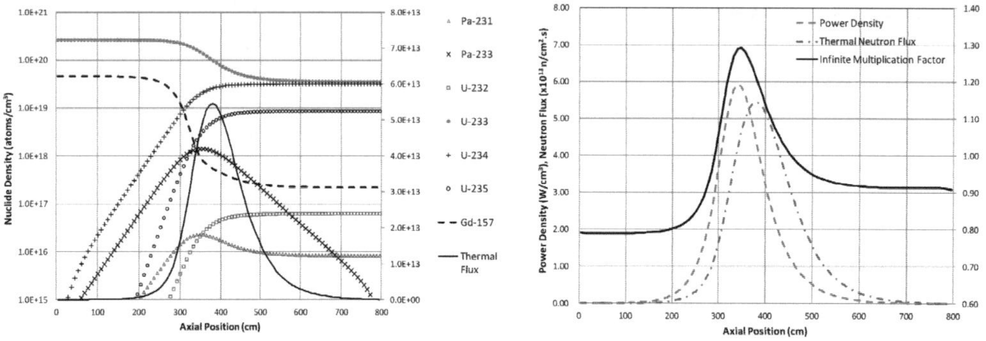

[Left] Nuclide densities (left axis) and thermal group neutron flux (right axis)
[Right] Power density, thermal group neutron flux (left axis) and kinf (right axis)
Figure 7. Critical and equilibrium condition of CANDLE HTGR with U-233 enrichment of 15 w/o and burnable poison Gd concentration of 8.75 w/o (reference case, JENDL-3.3).

Secondly, the impact of new libraries on the core characteristics of CANDLE HTGR are shown in Figs. 8 and 9, for uranium and thorium fuel cycles, respectively. The impact of the new libraries was shown in term of the ratio of the evaluated parameter (by a new library) divided by the one of the reference case (by JENDL-3.3 library).

The impact on the uranium fuel cycle case (shown in Fig. 8) is discussed first. For a particular fissile enrichment, the initial fuel composition (including Gd burnable poison) is identical however as can be observed in the Fig. 8, the kinf values show differences which systematically decrease as the fissile enrichment increases. For the kinf parameter, comparing with other new libraries, JENDL-4.0 based parameters are relatively closer to the reference case (JENDL-3.3). The equilibrium CANDLE HTGR effective multiplication factors (keff) of the new libraries show much better agreement with the reference keff (JENDL-3.3).

In a CANDLE reactor, a slower burning region velocity will result in a higher fuel burnup. These particular core characteristics can be observed from Fig. 8. Among the four values of fissile enrichment investigated in the present work, the fissile enrichment of 10 w/o shows the largest differences in the burning region velocity (approximately -2%) and consequently the achievable fuel burnup (approximately +2%). No clear systematic dependency of the velocity (and burnup) on the fissile enrichment. In addition, for this fissile enrichment (10 w/o), the maximum power density calculated by the new libraries shows also largest differences (approximately -3%) than other fissile enrichment does. Except for the initial fuel kinf, no significant differences are found among the new libraries for all parameters considered, i.e. their variations are less than ±1%.

Next, the impact on the thorium fuel cycle case (shown in Fig. 9) is discussed. In general, compared with the previous impact on the uranium fuel cycle case, except for the initial kinf, all parameters evaluated show much larger differences against the reference case as well as among the new libraries themselves.

Although the initial kinf differences against the reference case are smaller compared to the uranium case, the keff values at the equilibrium condition show relatively much larger differences for all fissile enrichment values considered (approximately +4%). Large differences against the reference case in the burning region velocity (and consequently the achievable fuel burnup) are also found for all fissile enrichment values considered, however, the ones of 6.5 w/o fissile enrichment show significantly large differences (approximately ±14%). For larger fissile enrichment values the differences are smaller, i.e. less than ±8%. The maximum power density differences of the new libraries against the reference case show a consistent dependency on the fissile enrichment, where the differences move from negative values (-8%) to positive values (+8%) as the fissile enrichment increases. The differences at most ±2% are found among the new libraries for all core parameters considered here.

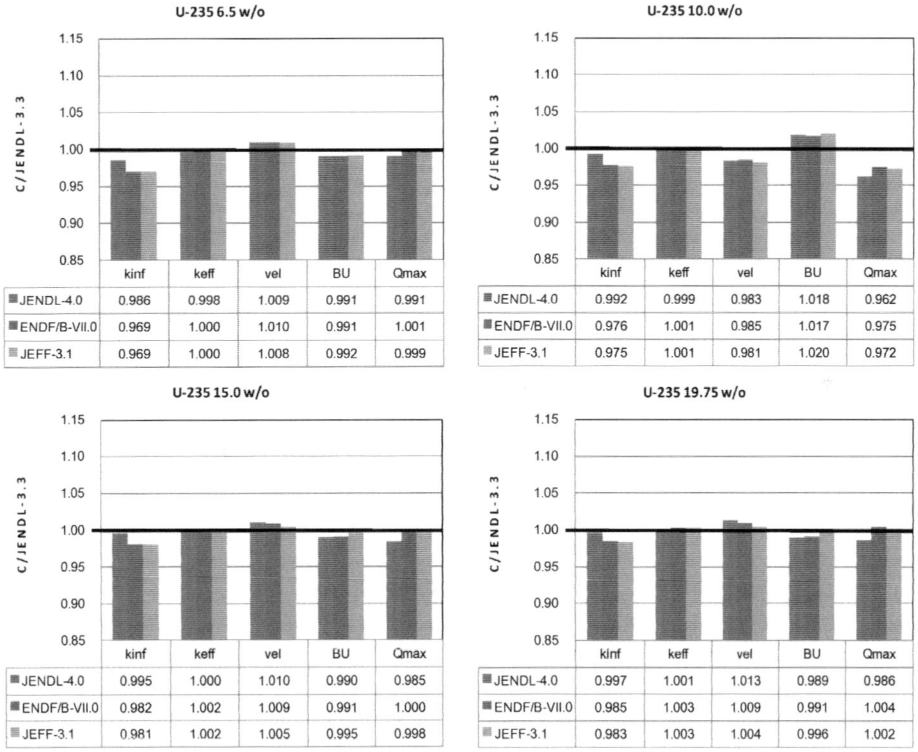

Figure 8. Impact of new libraries on the main parameters of the CANDLE HTGR core characteristics with uranium fuel cycle (ratio of a parameter calculated by new library and the one by JENDL-3.3).

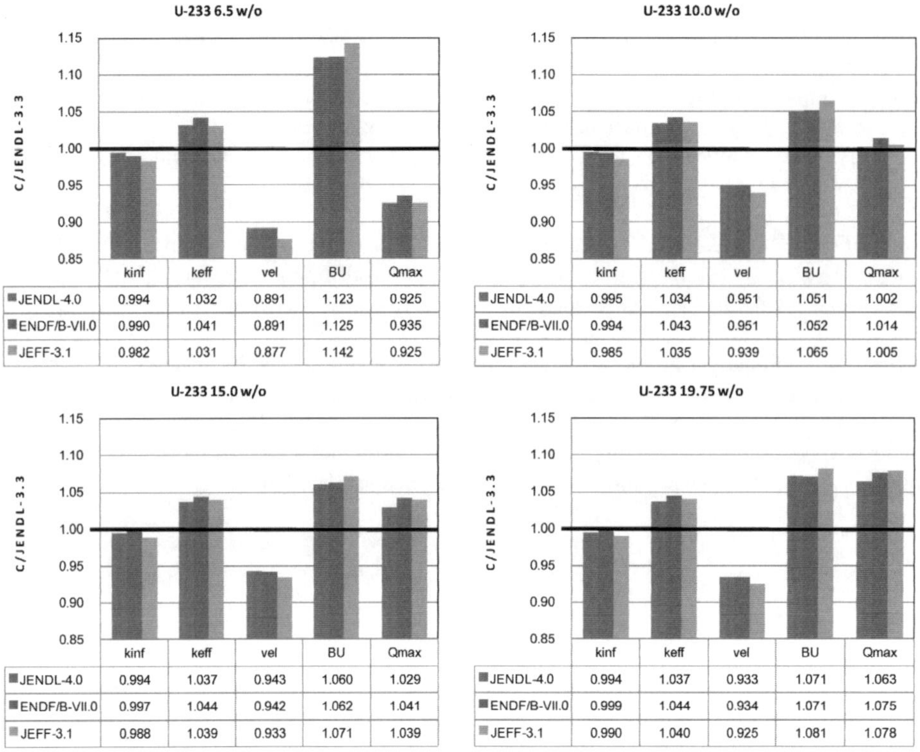

U-233 6.5 w/o

	kinf	keff	vel	BU	Qmax
JENDL-4.0	0.994	1.032	0.891	1.123	0.925
ENDF/B-VII.0	0.990	1.041	0.891	1.125	0.935
JEFF-3.1	0.982	1.031	0.877	1.142	0.925

U-233 10.0 w/o

	kinf	keff	vel	BU	Qmax
JENDL-4.0	0.995	1.034	0.951	1.051	1.002
ENDF/B-VII.0	0.994	1.043	0.951	1.052	1.014
JEFF-3.1	0.985	1.035	0.939	1.065	1.005

U-233 15.0 w/o

	kinf	keff	vel	BU	Qmax
JENDL-4.0	0.994	1.037	0.943	1.060	1.029
ENDF/B-VII.0	0.997	1.044	0.942	1.062	1.041
JEFF-3.1	0.988	1.039	0.933	1.071	1.039

U-233 19.75 w/o

	kinf	keff	vel	BU	Qmax
JENDL-4.0	0.994	1.037	0.933	1.071	1.063
ENDF/B-VII.0	0.999	1.044	0.934	1.071	1.075
JEFF-3.1	0.990	1.040	0.925	1.081	1.078

Figure 9. Impact of new libraries on the main parameters of the CANDLE HTGR core characteristics with thorium fuel cycle (ratio of a parameter calculated by new library and the one by JENDL-3.3).

4. Conclusion

The investigation on the impact of new evaluated nuclear data libraries against the core characteristics of small-sized, long-life CANDLE HTGRs shows that the new libraries have small impact on the uranium fuel cycle but have significant one on the thorium fuel cycle. In the future, more detail sensitivity analyses on the dominant nuclides and nuclear reaction types are needed to provide further feedback to the nuclear data community.

Acknowledgment

Some figures related to CANDLE burnup and cell calculation model used in the manuscript were provided by Drs. Ohoka Y. and Ismail.

References

[1] Sekimoto H, Ryu K and Yoshimura Y 2001 *Nucl. Sci. Eng.* **139** 306
[2] Shibata K, Iwamoto K, Nakagawa T, et al. 2011 *J. Nucl. Sci. Technol.* **48** 1
[3] Chadwick M B, Oblozinsky P, Herman M, et al. 2006 *Nucl. Data Sheets* **107** 2931
[4] Koning A, Forrest R, Kellet M, et al. 2006 *JEFF Report 21: The JEFF-3.0 Nuclear Data Library* OECD/NEA No. 6190
[5] Okumura K, Kugo T, Kaneko K and Tsuchihashi K 2007 *SRAC2006: A Comprehensive Neutronics Calculation Code System* JAEA-Data/Code 2007-004
[6] Ohoka Y and Sekimoto H 2004 *Nucl. Eng. Des.* **229** 15
[7] Ismail, Ohoka Y, Liem P H and Sekimoto H 2007 *Ann. Nucl. Energy* **34** 120
[8] Liem P H, Ismail and Sekimoto H 2008 *Prog. Nucl. Energ.* **50** 251

[9] Saito S, Tanaka T, Sudo Y et al. 1994 *Design of High Temperature Engineering Test Reactor (HTTR)* JAERI 1332
[10] Yamashita K, Shindo R, Murata I et al. 1996 *Nuc. Sci. Eng.* **22** 212
[11] Shibata K, Kawano T, Nakagawa T et al., 2002 *J. Nucl. Sci. Technol.* **39** 1125

Comparative Analysis on Nuclear Fuel Sustainability Aspect of FBR

Sidik Permana[1], Dwi Irwanto[1], Mitsutoshi Suzuki[2], Masaki Saito[3]

[1] Nuclear Physics and Bio Physics Research Group, Bandung Institute of Technology, Gedung Fisika, Jl. Ganesha 10, Bandung 40132, Indonesia.
[2] Department of Science and Technology for Nuclear Material Management (STNM), Japan Atomic Energy Agency (JAEA), 2-4 Shirane, Shirakata, Tokai Mura, Naka gun, Ibaraki 319-1195 Japan.
[3] Research Laboratory for Nuclear Reactors, Tokyo Institute of Technology 2-12-1 O-Okayama, Meguro-ku, Tokyo 152-8550, Japan

Email: psidik@fi.itb.ac.id

Abstract. Recycle program of spent nuclear fuel (SNF) will have some challanges in term of fuel cycle capability and its facilities as well as nuclear non-proliferation concern of special nuclear materials. A different analysis approach as a comparative study have been analyzed based on breeding ratio and heavy metal inventory ratio concepts in fast breeder reactor (FBR) type. Breeding ratio and heavy metal inventory obtain higher than unity which shows breeding gain or surplus inventory of heavy metals are obtained. Breeding ratio indicates the fuel conversion capability from conversion process of fertile materials into fissile material such as fertile materials of U-238, Pu-238, Pu-240 and fissile materials of Pu-239 and Pu-241. Inventory ratio approaches are appropriate to estimate some selected actinide as a mass inventory production such as plutonium inventory ratio which estimate the surplus mass inventory from the ratio of produced plutonium at the net of operation to the initial inventory ratio.

1. Introduction

As one of the important program to improve the reactor fuel utilization capability, fuel breeding program have been investigated in many years. Water cooled reactor technology uses uranium fuel as initial fuel to produce energy with lower fuel conversion ratio similar to high temperature reactor type [1-2]. To improve the fuel conversion capability, some modified reactor designs have been evaluated [3-4]. Recycling fuel program has some challenges to be optimized and harmonized for future nuclear fuel utilization suach as a reduction factor of the environmental effects, aspect of nuclear non-proliferation and vie point of nuclear fuel sustainability incuding recycling option of minor actinide (MA) actinides. Some SNF including MA can be recycled and some converted MA into plutonium can be considered to improve the level of nuclear nonproliferation [5-13]. Many purposes to pursue recycling program such as recycling transuranic programs which have been conducted for longer reactor operation purposes and to increase high burning MA level capability and transmutation of transuranium (TRU) in light water reactor (LWR) system and fast breeder reactor (FBR) system[14-

19]. A comparative analysis has been evaluated in the present study which is based on the breeding ratio and heavy metal inventory ratio approaches in fast breeder reactor (FBR) system.

2. Methods

Analysis of LWR fuel composition evaluation for different burnup or irradiation process and cooling time process have been investigated by adopting ORIGEN code [20]. Fast breeder reactor (FBR) core analysis was conducted by adoping a coupling codes of SLAROM, JOINT and CITATION codes based on nuclear data library of JFS-3-J-3.2R and JSFR (Japan Sodium Fast Reactor) as a basic FBR system [21-25]. As mentioned earlier that some composition of spent nuclear fuel (SNF) of LWR will be used as loaded intial fuel of FBR system.

2.1. Breeding Ratio and Heavey Metal Inventory Ratios

Sustainability Aspect of nuclear fuel in the reactors will be analyzed based on nuclear fuel breeding ratio and heavy metal inventory ratio. Fuel breeding analysis will be based on reaction rates of some fissile and fertile materials in the reactor such as uranium and plutonium fuel elements as main actinide. The equation to determine the breeding ratio is shown in **Eq. 1.**

$$BR = \frac{\left(\sum_{i=1}^{i=n} \Sigma_{c,i}^{fertile}\right)}{\left(\sum_{i=1}^{i=n} \Sigma_{a,i}^{fissile}\right)_v} \tag{1}$$

$\Sigma_{c,i}^{fertile}$: Macroscopic capture cross-section of fertile material for each i isotope

$\Sigma_{a,i}^{fissile}$: Macroscopic absorption cross-section of fissile materials for each i isotope

Heavy metal inventory ratio is used to estimate a mass inventory balance of heavy metals in the reactor such as plutonium and minor actinide (MA). Similar to analysis of plutonium inventory ratio, MA inventory ratio evaluation have been used for future transuranic mass inventory balance of FBR cycle [26]. Heavy metal inventory analysis is used to show the flow of nuclear fuel mass form supply stage up to discharged fuel. The ratio can be used to estimate the mass inventory balance based on the discharged fuel and supply fuel inventories.

$$IR_m = \frac{\left(\sum_{i=1}^{i=n} MI_{isotope\_m(i),disc}\right)}{\left(\sum_{i=1}^{i=n} MI_{isotope\_m(i),\sup}\right)} \tag{2}$$

IR_m : Mass Inventory Ratio of m-element

$MI_{isotope\_a(i),disc}$: Mass Inventory of isotope i of m-element at discharged fuel

$MI_{isotope\_a(i),\sup}$: Mass Inventory of isotope i of m-element at supply fuel

3. Results and Discussion

Some obtained results will be shown and discussed in this section including breeding ratio, and heavy metal inventory ratio during reactor operation. Fuel breeding ratio was studied to evaluate fuel conversion capability based on nuclear reaction aspect and to estimate some addition breeding gain. Spent nuclear fuels of LWR will be analyzed as initial fuels of FBR system based on LWR fuel burnup level of 50 GWd/t and cooling time process of 30 years. Plutonium composition of SNF LWR has even mass composition of plutonium such as Pu-238 and Pu-240 are about 30% and some odd mass plutonium are about 60% from total plutonium.

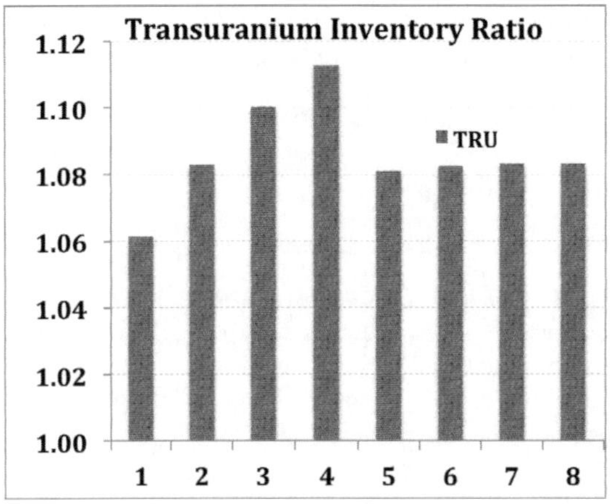

Figure 1 Transuranium Inventory Ratio of MOX fuel FBR.

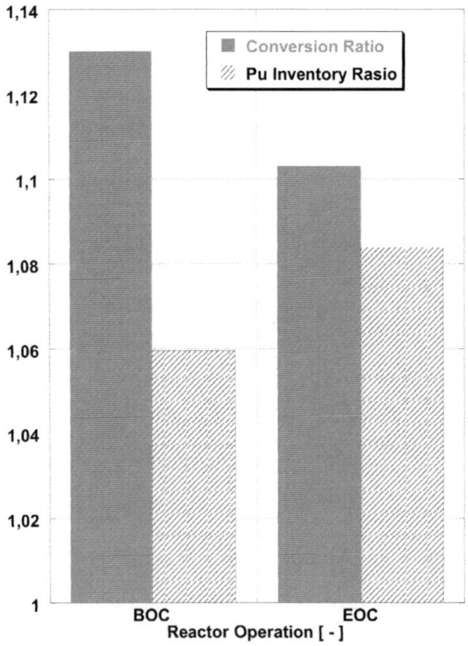

Figure 2 Fuel Conversion Ratio (Breeding Ratio) and Plutonium Inventory Ratio

3.1. Breeding and Heavy Metal Inventory Ratios

Obtained transuranium inventory ratio profile during reactor and fuel breeding ratio as well as plutonium inventory ratio are shown in Figures 1 and 2. Transuranium inventory ratio obtains more than unity during reactor operation as well as breeding ratio and plutonium inventory ratio. The results show that breeding gain level and heavy metal inventory gain of nuclear fuel can be estimated. The differences value of breeding gain and inventory gain are estimated because of the difference approach of achieving breeding gain. Heavy metal inventory ratio is only assuming the contribution of individual element or mixed element such as plutonium and transuranium element. Fuel breeding ratio uses main contribution materials such as uranium and plutonium as fertile and fissile materials.

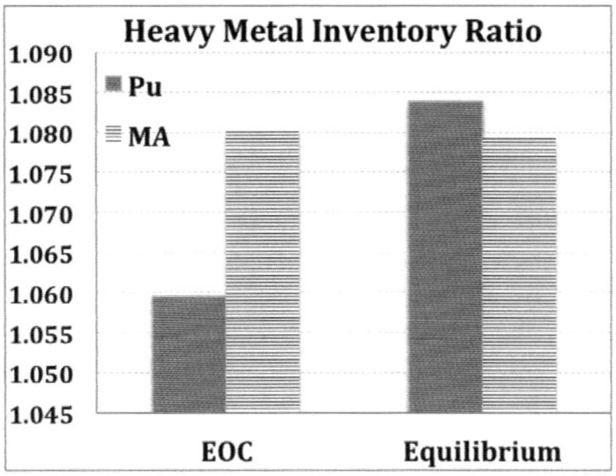

Figure 3 Heavy Metal Inventory of Plutonium and Minor actinide (MA).

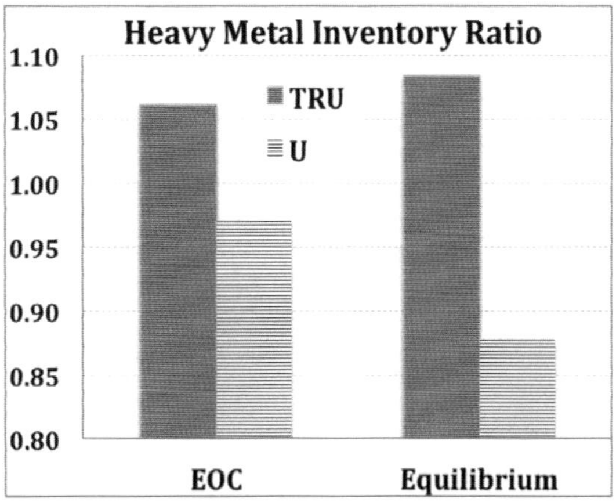

Figure 4 Heavy Metal Inventory Ratio of Transuranium (TRU) and Uranium

3.2. Transuranium Inventory Ratio

Heavy metal inventory ratios of FBR system for different reactor operation time are shown in Figures 3 and 4. All heavy metal element inventory ratio obtain higher than unity except uranium element which obtains less than unity. Uranium element inventory are less during reactor operation in comparing to the initial inventory ratio. It can be estimated that uranium element are reduced because of the fission reaction of U-235 and conversion process of U-238 into Pu-239 and so on. This is the reason why uranium inventory becomes less. On the other hand, transuranium element such as plutonium and minor actinide (MA) as well as transuranium element which is based on a mixed composition of plutonium and MA are always more than unity. More production of tranuranium elements are estimated during reactor operation in comparing to the initial composition. In equilibrium condition, plutonium element and transuranium element achieve higher inventory ratio than beginning of cycle (BOC) and minor actinide obtain less for equilibrium condition. This condition of MA can be estimated that during reactor operation, some Mas are converted into plutonium and it produced more plutonium by increasing the reactor operation and at the same time a reduction composition of MA is occur.

4. Conslusions

Breeding ratio and heavy metal inventory ratio approaches have been analyzed based on spent nuclear fuel of LWR as initial loaded fuel in fast breeder reactor (FBR) system. Transuranium and inventory ratio and fuel breeding ratio obtain more than unity during reactor operation. Breeding gain level and heavy metal inventory gain of nuclear fuel can be estimated. The differences value of breeding gain and inventory gain are estimated because of the difference approach of achieving breeding gain. All heavy metal element inventory ratio obtain higher than unity except uranium element which obtains less than unity because of the fission reaction of U-235 and fuel conversion process of U-238 into Pu-239 which causes uranium inventory becomes less. More production of transuranium elements are estimated during reactor operation in comparing to the initial composition. In equilibrium condition, plutonium element and transuranium element achieve higher inventory ratio than beginning of cycle (BOC) and minor actinide obtain less for equilibrium condition because of some MAs are converted into plutonium that makes a reduction composition of MA.

Acknowledgment

We would like acknowledge and extend our gratitude to desentralisasi research program of ministry of research, technology and higher education and ITB research innovation program for the research and international conference grant and publication.

References

[1] Duderstadt, J.J., Hamilton, L.J.,. Nuclear Reactor Analysis. Wiley, New York, (1976).
[2] P. Sidik, N. Takaki, H. Sekimoto, "Impact of different moderator ratios with light and heavy water cooled reactors in equilibrium states," *Ann. Nucl. Energy*, 33, 561 (2006).
[3] K. Hibi and H. Sekimoto, "Investigation of Neutron Reaction Behavior in Water-cooled FBR with MOX Fuel", *J. Nucl. Sci. Technol*, 42[2], 153 (2005).
[4] N. Takaki, "Neutronic Potential of Water Cooled Reactor with Actinide Closed Fuel Cycle", *Prog. Nucl. Energy*, 46[1-4], 223 (2000)
[5] M. Saito, et al., *Prog. in Nuc. Energy*, 40 (3-4), 365-374, 2002.
[6] International Atomic Energy Agency, Information Circular, INFCIRC/153, 1972.
[7] B. Pellaud, *J. Nucl. Mater. Managements*, XXXI(1),30-38, 2002.
[8] G. Kessler, *Nucl. Sci. and Eng*., 155, 53-73, 2007.
[9] P. Sidik, et al., *Nucl. Eng. and Des*., 241, 101-117, 2011.
[10] S Permana, M Suzuki, M Saito, Analysis on isotopic plutonium barrier based on spent nuclear fuel of LWR, Annals of Nuclear Energy 75, 116-122 (2015).
[11] S Permana, A Waris, Z Suud, M Suzuki, Analysis on Proliferation Resistance Factor and Fuel Breeding Capability Based on Even Mass Plutonium Isotope Compositions, Energy Procedia, 71, 182-190 (2015).

[12] S Permana, M Suzuki, M Saito, A Waris, Z Suud, Study on material attractiveness aspect of spent nuclear fuel of LWR and FBR cycles based on isotopic plutonium production, Energy Conversion and Management, 72, 19-26, (2013).

[13] Sidik Permana, Mitsutoshi Suzuki, Basic evaluation on material attractiveness of isotopic plutonium barrier, Progress in Nuclear Energy, 53[7], 958 (2011).

[14] Sidik Permana., Zaki, Suud, Core Performance and Plutonium Production of Small Long Life Fast Reactor using Doping Actinides, Proceeding of International Congress on Advance In Nuclear Power Plants (ICAPP), Tokyo-Japan, 2009.

[15] Sidik Permana, Zaki Suud, and Mitsutoshi Suzuki, Trans-uranium doping utilization for increasing protected plutonium proliferation of small long life reactor, Proceedings of Global 2009, "The Nuclear Fuel Cycle: Sustainable Options & Industrial Perspectives", Paris, France, 2009.

[16] H. Choi and T. J. Downar, 1999, A Liquid-Metal Reactor for Burning Minor Actinides of Spent Light Water Reactor Fuel—I: Neutronics Design Study, Nuclear Engineering and Design, 133, 1–22,1999.

[17] Edward A. Hoffman and Weston M. Stacey, Comparative Fuel Cycle Analysis of Critical and Subcritical Fast Reactor Transmutation Systems, Nuclear Technology 144, 83, 2003.

[18] T. A. Taiwo, et al., Assessment of a Heterogeneous PWR Assembly for Plutonium and Minor Actinide Recycle, Nuclear Technology 155, 34, 2006.

[19] S Permana., et al , Reactor core and actinide production evaluation based on different loading material of recyled spent nuclear fuel of LWR in FBR, International journal of hydrogen energy 41, 7076 (2016).

[20] Oak Ridge Isotope GENeration Code, "ORIGEN-2, Isotope Generation and Depletion Code", ORNL TM-7175, July,1980.

[21] M. Nakagawa, K. Tsuchihashi, "SLAROM: A Code for Cell Homogenizations Calculation of Fast Reactor, JAERI 1294", Japan Atomic Energy Research Institute, 1984.

[22] T. B. Fowler, D. R. Vondy, G. W. Cunningham, "Nuclear Reactor Core Analysis Code: CITATION, Oak Ridge National Laboratory report", ORNL-TM-2496, Rev2, USA, 1971.

[23] G. Chiba et al., *J. At. Energy Soc. Jpn.*, 14, 335, 2002.

[24] T. Nakagawa *et al.*, *J. Nucl. Sci. Technol.*, 32, 1259, 1995.

[25] S. Ohki, et al., "Design Study of Minor Actinide Bearing Oxide Fuel Core for Homogeneous TRU Recycling Fast Reactor System", *The 10th OECD/NEA P&T Meeting,* Mito, Japan, 6-10 October, 2008.

[26] T. Yokoo, T. Inoue, "The transuranic mass balance during the introduction of metal fuel FBR cycle, "*Nucl. Sci. Technol.*, 36[9], 761 (1999).

Application Electrochemical Impedance Spectroscopy Methods to Evaluation Corrosion Behavior of Stainless steels 304 in Nanofluids Media

Djoko Hadi Prajitno[1], Efrizon Umar[1] and Dani Gustaman Syarif[1]

[1]PSTNT-BATAN Jl. Tamansari 71 Bandung 40132, Indonesia

Email: djokohp@batan.go.id

Abstract. Corrosion is a common problem in many engineering metals and alloys. Electrochemical methods are commonly instrument to use as tool to study the corrosion behavior of the metals and alloy. This method was examined interaction between a surface of the metals and alloys in corrosive media. The present paper, the effects of nano particle ZrO2 as an additive to aqua de mineralized on the corrosion behavior of stainless steel were investigated. Electrochemical impedance spectroscopy (EIS) testing was performed in both de mineralized water and demineralized water contain nano particle 0,01% ZrO2 as Nano fluid. Surface morphology examination of the specimens showed that microstructure of stainless steel 304 alloys relatively unchanged after corrosion and EIS testing. According to the corrosion potential examination of the stainless steel 304 in nanofluid media, it showed that stainless steel 304 actively corroded in nanofluida media. The value of anodic Tafel slope stainless steel 304 in nanofluid higher compare with in demineralized water. Tafel polarization examination show that corrosion rate of stainless steel 304 in nanofluid higher compare with corrosin rate in demineralized media.EIS technique show that impedance of stainless steel 304 in nanofluid lower compare with in demineralized media, resulting in an increase in the corrosion rates of these stainless steel 304 specimens in nano fluids

1. Introduction

Electrochemical impedance spectroscopy (EIS) is a technique commonly used for studying corrosion of metal and alloy in environment system. A majority of the experiments apply small amplitude perturbation to measure the impedance as a function of frequency and the analyses employ linearization of the equations [1,2]. EIS is non destructive testing to observe changes in the surface electrode as a function with time. Basically, EIS measures the electrochemical response to a minute AC voltage applied at different number frequencies. This electrochemical reaction occured on the surface of anode and cathode electrode is interpreted in terms of an equivalent circuit [3,4]. For example, a capacitor has the same frequency response as a reaction step when electrons or ions build up on a surface, and a resistor represent the transport of charge surface of materials or interface between solution and surface of metal. Results are commonly reported as a plot of imaginary vs. real components of impedance as a Nyquist plots or impedance vs. frequency and phase angle vs. frequency as a Bode plots [5,6,7].

The present work is to investigate the electrochemical impedance spectroscopy (EIS) behavior of stainless steel 304 in Nano fluid solution contain 0,1 gpl nanoparticle ZrO2, A series of experiments have been conducted, including potential measurement to assess the influence of nanoparticles against on open circuit potential (OCP) and Tafel polarization measurements to study the corrosion of the stainless steel 304 and EIS.

2. Experimental Methods

2.1. Materials and nanofluids

The stainless steel 304 used in this experiments were discs stainless having 1,6 Cm in diameter and 2 mm in thickness. The chemical composition (in wt%) 18,21 % Cr, 8,40% Ni, 0,39% Si, 1,56% Mn, 0,45%Mo, 0,025% C balanced Fe. The specimen was wet abraded with 500, 800 and 1200 emery paper, degreased with detergent finally cleaned with ultrasonic cleaner in alcohol solution [8]. Nanofluid preparation were described in [9]

2.2. EIS study

Electrochemical experiments were conducted in three electrode cell, a stainless steels 304 disk as the working electrode, a saturated calomel electrode (SCE) as reference electrode and platinum wire were used as an auxiliary electrodes. The experiments were conducted by an instrument PCI4, G300 Potentiostat/Galvanostat series and connected to a personal computer provide with DC 105 software from Gamry [10]. Electrochemical impedance spectroscopy curve were obtained by the polarization of working electrode from 100 KHz to 10 Hz with ac current 0.1mA. Nyquist and Bode plot were acquired by EIS300 software from Gamry.

2.3. Characterization

Corroded and un corroded samples were analyzed by optical microscopy.

3. Result and Discussions

Figure 1 shows the variation of open circuit potential (OCP) with immersion time 600S of stainless steel 304 alloys in demineralized water and demineralized water + 0,01gpl ZrO2 Nano fluid at the room temperature. At the primary stage, OCP of stainless steel 304 alloys in demineralized water and demineralized water + 0,01gpl ZrO2 Nano fluid at the value almost of -38 mv then decreases rapidly. After 20S the value of OCP SS 304 in demineralized water and demineralized water + 0,01gpl ZrO2 Nano fluid nearly -88 mV and -119 mV respectively. After that the variation of OCP with time are increases slowly with SS 304 in demineralized water lower compared with demineralized water + 0,01gpl ZrO2 Nano fluid all the time. Finally, after 600S the value of SS 304 in demineralized water and demineralized water + 0,01gpl ZrO2 Nano fluid were -65mV and -126 mV respectively.

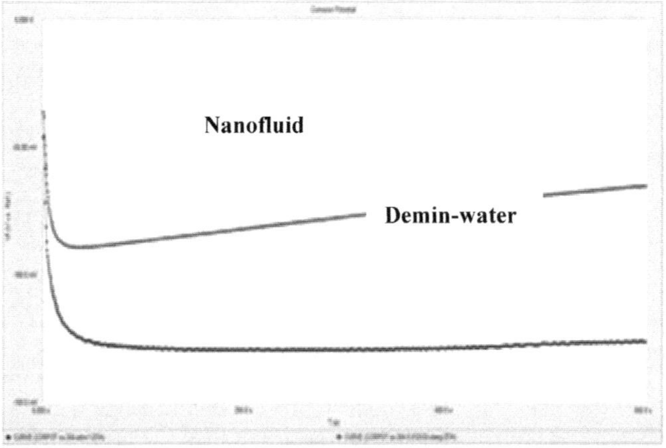

Figure 1. The OCP of stainless steel 304 immersed in demineralized water and Nano fluid as a function of time

3.1. This is indicates that the OCP of stainless steel 304 in Nano fluids more positive compared with demineralized water due to corrosion of stainless steels 304 before stable protective film took place. The OCP of demineralized water + 0,01gpl ZrO2 Nano fluid higher compared with demineralized water indicates nano fluid more corrosive compared with demineralized water.

Polarization curve of stainless steel 304 in demineralized water and demineralized water + 0,01gpl ZrO2 Nano fluid showed in figure 2. It should be noticed that OCP and the anodic and cathodic Tafel constants value of stainless steel 304 are depend on electrolyte solution. From figure 2 it can be presented that OCP stainless steel 304 in demineralized water + 0,01gpl ZrO2 nano fluid more positive (-83 mv) compared with demineralized water (-134mv). This result indicated that potential stainless steel 304 in demineralized water + 0,01gpl ZrO2 Nano fluid more anodic and nano fluid more corrosive compared with demineralized water. The anodic and cathodic Tafel constant of stainless steel 304 alloys in demineralized water and demineralized water + 0,01gpl ZrO2 Nano fluid displayed the kinetic anodic and cathodic reaction. The anodic and cathodic curve of stainless steel 304 alloys in demineralized water revealed smoothly compare with in demineralized water + 0,01gpl ZrO2 Nano fluid. The anodic and cathodic plateau of stainless steel 304 in demineralized water + 0,01gpl ZrO2 nano fluid is highest compared with demineralized water. This results presented that corrosion stainless steel 304 in demineralized water + 0,01gpl ZrO2 nanofluid more dynamic of charge transfer and resistance of nanofluid during equilibrium. From Figure 2, anodic and cathodic Tafel constant and corrosion rate of stainless steel 304 in demineralized water and demineralized water + 0,01gpl ZrO2 nano fluid can be calculated by using DC 105 corrosion software from Gamry.

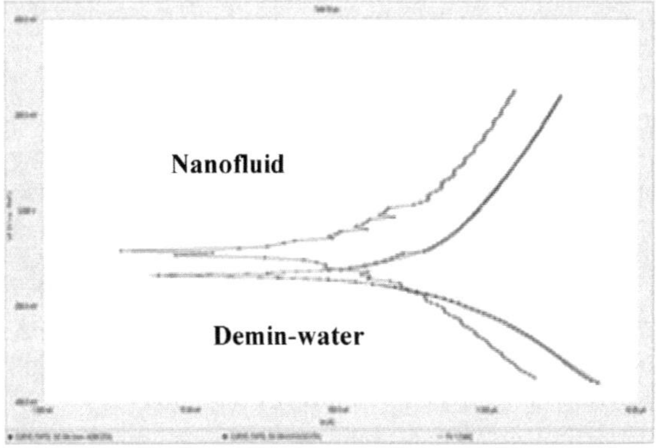

Figure 2. Tafel polarization curve stainless steel 304 immersed in de mineralized water and Nano fluid

3.2. Cathodic and anodic Tafel constants value of stainless steel in demineralized water and Nano fluid are 0,188v/decade and 0,249v/decade and 0,227 v/decade and 0,315 v/decade, respectively. Moreover corrosion rate stainless steel 304 in demineralized water and demineralized water + 0,01gpl ZrO2 Nano fluid are 0,058 MPY and 0.069 MPY, respectively.

Electrochemical impedance spectroscopy (EIS) stainless steel 304 were performed in demineralized water and demineralized water + 0,01gpl ZrO2 Nano fluid in order to evaluate electrochemical behavior of stainless steel 304 in different electrolyte. EIS data was recorded for all two of the electrolytes demineralized water and demineralized water + 0,01gpl ZrO2 Nano fluid after immersion of stainless steel 304 up to 10 minute. The Nyquist plot of data real impedance as vertical axis and imaginary impedance as horizontal axis showed in Fig.3. Fig. 3 shows that impedance increases with time for stainless steel 304 in all two electrolyte demineralized water and demineralized water + 0,01gpl ZrO2 Nano fluid. The results show that the impedance decreases as demineralized water content nano particle ZrO2. Decreases impedance suggesting an activation of the corrosion process took place in stainless steel 304 in the presence of nano particle ZrO2 of demineralized water. This results in good agreement with the Tafel polarization curves where an enhancement of the corrosion occurred if stainless steel 304 immerse in nanofluid contain nano particle ZrO2.

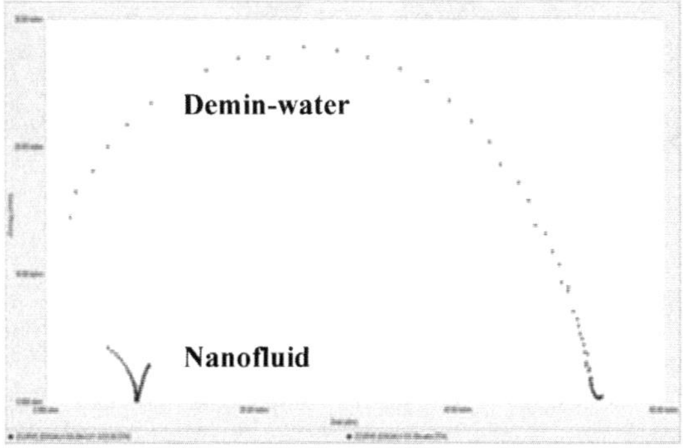

Figure 3. Nyquist plot stainless steel in demineralized water and Nano fluid

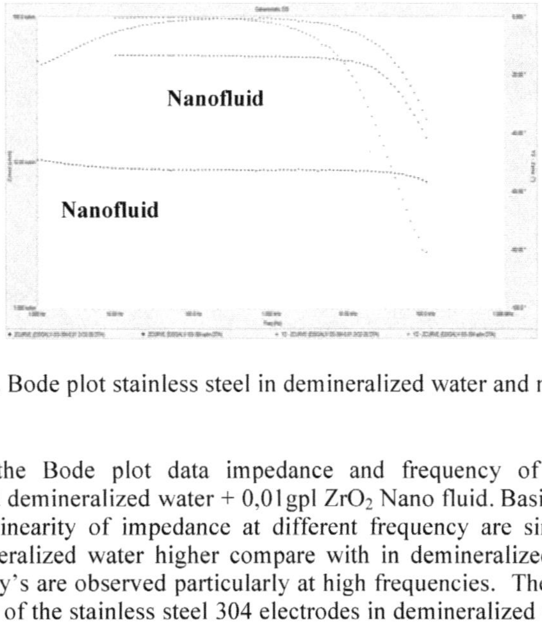

Figure 4. Bode plot stainless steel in demineralized water and nano fluid

Figure 4 depicted the Bode plot data impedance and frequency of stainless steel 304 in demineralized water and demineralized water + 0,01gpl ZrO_2 Nano fluid. Basically, it can be observed that the conditions of linearity of impedance at different frequency are similar but impedance of stainless steel in demineralized water higher compare with in demineralized water + 0,01gpl ZrO_2 Nano fluid. Non-linearity's are observed particularly at high frequencies. The Bode plots are suitable to compare the behavior of the stainless steel 304 electrodes in demineralized water and demineralized water + 0,01gpl ZrO_2 Nano fluid due to the electrochemical reaction. The large changes in impedance took place because the redox reaction on the stainless steel in demineralized water and demineralized water + 0,01gpl ZrO_2 Nano fluid is different in the electron transfer. The effect of the charge transfer affect the corrosion rate stainless steel in demineralized water lower compare with in demineralized water + 0,01gpl ZrO_2 Nano fluid.

Surface morphology of specimens stainless steel 304 before and after corrosion testing in demineralized water and Nano fluids showed in Figure 5 a and b, respectively. In Figure 5a and 5b showed that the surface morphology specimen still free of product corrosion scale. From the data surface morphology stainless steel, it can be concluded that Nano fluid solution is not corrosive electrolyte.

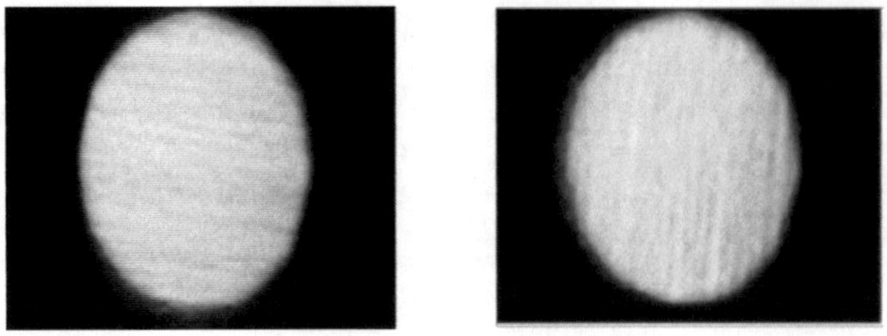

a b

Figure 5. Surface morphology of stainless steel before (5a) and after (5b) corrosion test in the Nano fluids

4. Conclusion

The corrosion stainless steel in nanofluid contains adm+0.01 gpl ZrO_2 nanoparticles was performed by Tafel polarization and Electrochemical Impedance spectroscopy. Following results are obtained

1. The results show that according corrosion potential examination of the stainless steel 304 in nanofluid media 0.01gpl ZrO2 nanoparticle was actively corroded.
2. The value of anodic Tafel slope stainless steel 304 in nanofluid higher compare with in demineralized water.
3. The results of Tafel polarization technique show that corrosion rate of stainless steel 304 in nanofluid higher compare with corrosin rate in demineralized media.
4. The results of EIS technique show that impedance of stainless steel 304 in nanofluid lower compare with in demineralized media.

Acknowledgments
This work was long term research program in corrosion of metals in nanofluids media and financially supported by the DIPA Program 2015 from Center for Applied nuclear science and technology–National Nuclear Energy Agency.

References

[1] S. Noyel Victoria, S. Ramanathan, Effect of potential drifts and ac amplitude on the electrochemical impedance spectra, Electrochimica Acta 56 (2011) 2606–2615
[2] M. Orazem, B. Tribollet, Electrochemical Impedance Spectroscopy, John Wiley and Sons, NJ, 2008.
[3] Samue l. Zelinka,at all, Electrochemical impedance spectroscopy (EIS) as a tool for measuring corrosion of polimer-coated fastener used in treated wood,Journal Forest Product, vol.59, No.1/2 (2009) 77.
[4]. G. Rondellia, al all. Electrochemical In vitro corrosion study by EIS of a nickel-free stainless steel for orthopaedic applications, Biomaterials 26 (2005) 739
[5] L. Freire, M.J.Carmezima, M.G.S.Ferreiraa, M.F.Montemora, The electrochemical behavior of stainless steel AISI304 in alkaline solutions with different pH in the presence of chlorides Electrochimica Acta 56 (2011) 5280
[6] Farzad Mohammadia, Tirdad Nickchia, M.M. Attara,Akram Alfantazia EIS study of potentiostatically formed passive film on 304 stainless steel, Electrochimica Acta 56 (2011) 8727
[7] F. Louis Floyda, at all, Using electrochemical impedance spectroscopy to predict the corrosion resistance of unexposed coated metal panels, Progress in Organic Coatings 66 (2009) 8

[8] Djoko Hadi Prajitno, Effects of scan rate on the corrosion behavior SS 304 stainless steel in the nanofluid measured by Tafel polarization methods, AIP Conference Proceedings 1677, 070011 (2015);
[9] Djoko Hadi Prajitno and Dani Gustaman Syarif, Corrosion of Carbon Steel in Nanofluid Containing ZrO2 Nanoparticle at Different Temperature Advanced Materials Research Vol. 896 (2014) pp 168-172
[10] Gamry Potentiostat/Galvanostat CI4/G300, Electrochemical Measurement System 2010 and Gamry software DC 105 corrosion analysis system 2010

The prediction of helium gas viscosity under high pressure and high temperature with the Chapman-Enskog solution and excess viscosity

Elin Yusibani[1*], Yasuyuki Takata[2], Zaki Suud[3] and Dwi Irwanto[3]
[1]*Physic Department, FMIPA, Universitas Syiah Kuala, Banda Aceh 23111, Indonesia*
[2]*Department of Mechanical Engineering, Kyushu University, 744 Motooka Nishiku, 819-0395, Fukuoka, Japan*
[3]*Physic Department, Faculty of Mathematic and Natural Sciences, Bandung Institute of Technology, Ganeca 10 Bandung, Indonesia*

Email: e_yusibani@unsyiah.ac.id

Abstract. The purpose of this work is to predict a helium gas viscosity under high pressure and high temperature for practical industrial uses. The suitable force constants and a collision integral for the Chapman-Enskog solution to estimate viscosity in the limit of zero density were recommended by the present author. At high density, modification of the Arp and McCarty extrapolation equation for excess viscosity was applied. A combination of the Chapman-Enskog solution and modification of the Arp and McCarty excess viscosity gives an estimation of helium gas viscosity within 2 to 5 % deviation from the existing experimental data under high-temperature and high-pressure region.

1. Introduction

Helium (Helium-4) is a noble gas that mostly uses in the industrial purpose. There is a need for recommendations relating to the best available viscosity data and the best available correlations for practical purposes under high pressure and high temperature for helium economy. Especially in a nuclear power plant generation IV, Helium is often used as a coolant in High Temperature Gas-cooled Reactors (HTGR), it is one among the other type of nuclear reactor for Generation IV (Gen-IV) of advance nuclear reactors system [1,2]. At atmospheric pressure, helium gas viscosity may be considered to be at the limit of zero density, $\eta_0(T)$, which can be well represented by the Chapman-Enskog solution as a first approximation [3]. Besides the Chapman-Enskog solution, empirical correlations are also can be applied. However, the latest commonly represented in a complicated equation and not reccommended. The present author has been succeded to predict hydrogen gas viscosity under high pressure and high temperatur by the Chapman-Enskog solution and the excess viscosity within 2 to 4 % of deviations [4]. The same method with previous work will be applied for helium.

2. Existing experimental data

Helium is the second lightest after hydrogen. Moreover helium is a colorless, odorless and tasteless. It introduced difficulties to be measure in an accurate way due to leakage. There are several experimental techniques available to obtain helium gas viscosity data. **Table 1** show the selected experimental data which are used to be compared with the present correlation. The highest temperature of about 900 K obtained by Makavetskas with the capillary method. The highes pressure of 100 MPa have done by team of Hydrogenius in Japan with the same method with Makavetskas. Measuring viscosity at high pressure and temperature is often measured with capillary tube.

Table 1. Selected experimental data set

References/Methods	Uncer./%	T/K	P/MPa
Seibt 2009 [5]/Vibrating-wire	0.25 to 0.3	293	up to 30
Ross (1957) [6]/Capillary tube	1	*298* to 323	3,5 to 68,9
Gracki (1969) [7]/Capillary tube	0.1 – 0.2	-373 to 298	1,5 to17,5
Trauts (1930) [8]/Capillary tube	2	293 to 523	0.1
Makavetskas (1963)[9]/Capillary tube	-	283 to 918	5-51
Enrique (1966)[10]/transpiration	ns	373 to *473*	up *to* 82
Kao&Kobayashi (1967)[11]/Capillary tube	0.15	183 to 323	1 to 50
Tsederberg (1969)[12]/Capillary tube	3	292	5 to 29
Kestin&Leidenforst (1959)[13]/Oscillation disk	0.2 – 0.5	298 to 510	0,1 to 13,7
Evers (2002) [14]/Oscillating disk and oscillating wire	0.6-1.0	233 to523	Up to 30
Flynn (1963)[15]/Capillary flow	0.1	223 to 523	Up to 30
HYDROGENIUS[16]/Capillary method (2012)	2	295-500	Up to 100

3.The Enskoq theory

We can have assumptions that a gas can be treated as an ensemble of hard elastic spheres and multiple collisions can be neglected. ψ is the collisional transfer of molecular property across an element of area dS at point r. Then the collisional part of the vector of flow of ψ is found to be

$$\frac{1}{2}\chi\sigma^3 \iiint \mathrm{m}(\psi' - \psi)\mathrm{ff}_1\, k(g.k)dkdcdc_1 + \frac{1}{4}\chi\sigma^4 \iiint (\psi' - \psi)ff_1\, k.\nabla\ln(f/f_1)\mathrm{k}(g.k)dkdcdc_1 \quad (1)$$

Where all the quantities are evaluated at the point r since the second term of this already involves space-derivatives, f and f_1 can here be replaced by $f^{(0)}$ and $f_1^{(0)}$. Consider first the transfer of momentum. Let

$$\psi = mC = m(c - c_0) \quad (2)$$

Where c_0 is evaluated at the special point r under consideration, and does not vary with the position of the molecule, the collision part becomes

$$\frac{1}{2}\chi\sigma^3 \iiint \mathrm{m}(\mathrm{C}' - \mathrm{C})\mathrm{ff}_1\, k(g.k)dkdcdc_1 + \frac{1}{4}\chi\sigma^4 \iiint \mathrm{m}(\mathrm{C}' - \mathrm{C})\mathrm{f}^{(0)}\mathrm{f}_1^{(0)}\, k.\nabla\ln(\mathrm{f}^{(0)}/\mathrm{f}_1^{(0)})\mathrm{k}(g.k)dkdcdc_1 \quad (3)$$

We use the relation $f = f^{(0)} \equiv n\left(\frac{m}{2\pi kT}\right)^{3/2} e^{-mC^2/2kT}$, after integration with respect to k, using

$$\int(\mathrm{C}' - \mathrm{C})\,\mathrm{k}(\mathrm{g.k})\mathrm{dk} = \int \mathrm{kk}(\mathrm{g.k})^2\mathrm{dk} = \frac{2\pi}{15}(2\mathrm{gg} + \mathrm{g}^2\mathrm{U}) \quad (4)$$

$$\int(\mathrm{C}' - \mathrm{C})\,\mathrm{k}(\mathrm{k.v})(\mathrm{g.k})\mathrm{dk} = \int \mathrm{kk}(\mathrm{v.k})(\mathrm{g.k})^2\mathrm{dk} = \frac{\pi}{12}\left[\mathrm{v.g}\left(\frac{(\mathrm{gg}+\mathrm{g}^2\mathrm{U})}{\mathrm{g}}\right) + \mathrm{g}(\mathrm{vg} + \mathrm{gv})\right] \quad (5)$$

The expression becomes

$$\frac{\pi}{15}\chi\sigma^3 m \iint ff_1\,(2gg + Ug^2)dcdc_1 + \frac{\pi}{48}\chi\sigma^4 m \iint f^{(0)}f_1^{(0)}\left[\left\{g.\nabla\ln\left(\frac{f^{(0)}}{f_1^{(0)}}\right)\right\}\frac{(gg+g^2U)}{g} + g\left\{g\nabla\ln\left(\frac{f^{(0)}}{f_1^{(0)}}\right) + \nabla\ln\left(\frac{f^{(0)}}{f_1^{(0)}}\right)g\right\}\right]dcdc_1 \qquad (6)$$

In the first integral, since for any function φ

$$\int f\phi dc = \int f_1\phi_1 dc_1 = n\left(\frac{m}{2\pi kT}\right)^{3/2}\int \phi e^{-mC^2/2kT}dC = n\bar{\phi} \qquad (7)$$

and $g = C_1 - C$, when $\overline{C_1} = 0, \bar{C} = 0$, The value of this integral is

$$\frac{1}{5}b\rho^2\chi(2\overline{CC} + \overline{C^2}U) \qquad (8)$$

Where $b=2\pi\sigma^3/3m$ and $\rho = nm$. In the second integral, since

$$\nabla\ln\left(\frac{f^{(0)}}{f_1^{(0)}}\right) = \frac{m}{2kT^2}(C^2 - C_1^2)\nabla T + \frac{m}{kT}(\nabla c_0)(C - C_1) \qquad (9)$$

In terms involving ∇T yield odd functions of C, C_1, which vanish on integration; hence the integral is equal to

$$-\frac{\pi}{48kt}\chi\sigma^4 m^2 \iint f^{(0)}f_1^{(0)}\left[(\nabla c_0\colon gg)\frac{(gg+g^2U)}{g} + g\{(\nabla c_0.g)g + g(\nabla c_0.g)\}\right]dcdc_1 \qquad (10)$$

This can readily be evaluated by charging the variable of integration from c and c_1 to G_0 and g, where $G_0=1/2(C+C_1)$. Its value is found to be

$$-\varpi\left\{\tfrac{6}{5}\overline{\overline{\nabla c_0}} + U\nabla.c_0\right\} = -\varpi\left\{\tfrac{6}{5}\dot{e} + U\nabla.c_0\right\} \qquad (11)$$

Where

$$\varpi = \tfrac{4}{9}\pi^{\frac{1}{2}}\chi n^2\sigma^4\sqrt{mkT} = \frac{b^2\rho^2 x\sqrt{mkT}}{\pi^{3/2}\sigma^2} \qquad (12)$$

Alternatively, in term of the coefficient of viscosity at ordinary pressure namely

$$\mu = \frac{1.016 \times 5\sqrt{mkT}}{16\sqrt{\pi}\sigma^2}$$

ϖ is given by

$$\varpi = 1.002\mu\chi b^2\rho^2 \qquad (13)$$

Adding (8) and (11), we obtain for the value of (7)

$$\frac{1}{5}b\rho^2\chi\left(2\overline{CC} + \overline{C^2}U\right) - \varpi\left\{\tfrac{6}{5}\dot{e} + U\nabla.c_0\right\} \qquad (14)$$

This gives the part of the pressure tensor which arises from the collision transport of momentum. To do this we must add $\rho\overline{CC}$, which gives the rate of transport by molecular motions; then the whole pressure tensor is found to be

$$P = \rho\left(1 + \tfrac{2}{5}b\rho\chi\right)\overline{CC} + \frac{1}{5}b\rho^2\chi\overline{C^2}U - \varpi\left\{\tfrac{6}{5}\dot{e} + U\nabla.c_0\right\} \tag{15}$$

on substituting in this the value of $\rho\overline{CC} = knTU$, Where $p_0 = knT(1 + b\rho\chi)$ it becomes

$$P = (p_0 - \varpi\nabla.c_0)U - \rho\left\{2\mu\chi^{-1}(1 + \tfrac{2}{5}b\rho\chi)^2 + \tfrac{6}{5}\varpi\right\}\dot{e} \tag{16}$$

For a gas in a uniform steady state, *P* reduces to a hydrostatic pressure p_0- A result already implicit. For a gas not in a uniform steady state, the hydrostatic pressure is $p_0 - \varpi\nabla.c_0$. The extra term $-\varpi\nabla.c_0$ differs from zero whenever the density of the gas is varying; it represents a volume viscosity similar to that opposing expansion or contracting of the gas. The deviation of the pressure system from the hydrostatic is given in last term of (16), which represents the usual shear viscosity. Equating this term to -2μ\`e, we see that in a dense gas the coefficient of viscosity μ\` is given in term μ, the viscosity at ordinary densities (at the same temperature), by the relation

$$\mu` = \mu\chi^{-1}\left(1 + \tfrac{2}{5}b\rho\chi\right)^2 + \tfrac{3}{5}\varpi$$
$$\mu = \mu_0 b\rho\left(\frac{1}{b\rho x} + 0.800 + 0.7614b\rho x\right) \tag{17}$$

The function η/ρ has a minimum value for *bρx*=1,146, The values *bρ* and *bρx* are obtained from the comparison of the experimental P-V-T data with Enskog equation of state

$$P + a\rho^2 = \frac{R\rho T(1+b\rho x)}{M} \tag{18}$$

Enskog suggested *x* and *b* to be independent of temperature. The first assumption is maintained, the latter however disregarded. With the substitutions $v_A = 1/\rho_A$; $\rho_A = \rho\rho_0$; $a` = a\rho_0^2$ $b` = b\rho_0$ Equation (18) reads

$$Pv_A = \frac{R\rho_0 T}{M} + \frac{R\rho_0 Tb`\rho_A x}{M} - a`\rho_A \tag{19}$$

This equation (19) can be compared with the virial equation of state

$$Pv_A = A + B\rho_A + C\rho_A^2 + \cdots \tag{20}$$

gives

$$PV_A = A + Ab`\rho_A x - a`\rho_A \tag{21}$$

Since *x* approaches unity if the density approaches zero, resulting in the relations

$$A = \frac{R\rho_0 T}{M} \tag{22}$$

$$Ab` - a` = B \tag{23}$$

And a possible transcription of (19) as

$$Pv_A = A + B\rho_A + Ab`(x-1)\rho_A \tag{24}$$

IOP Conf. Series: Journal of Physics: Conf. Series **799** (2017) 012008 doi:10.1088/1742-6596/799/1/012008

With the calculation of the values *bρx* from equation (14), the coefficients of viscosity can be determined with equation (8) and the experimental value for μ_0. Finally it is possible to calculate a value for the diameter of the molecule with the expression, given by Enskog [17,18,19,20]

$$\sigma = \frac{1}{0.955}\sqrt{M/RT}(\eta/\rho)_{min} \tag{25}$$

To estimate helium gas viscosity, we can predict at the limit of zero density, which can be well represented by the Chapman-Enskog solution as a first approximation

$$\eta_o(T) = \frac{5}{16}\frac{\sqrt{\pi mkT}}{\pi\sigma^2\Omega_\eta(T^*)} \tag{26}$$

The relation need three parameters i.e. collision diameter, collision integral as a function of reduced temperature. **Table 2** shows various force constants and collision integrals for helium gas.

Table 2. Intermolecular force constants for helium gas

Reference	**Force constant**	
	σ (nm)	**ε/k (K)**
Mason (Exp-six)[21]	0,3135	9,16
Mason (L-J)	0,2869	10,22
Saxena (L-J)[22]	0,2556	10,22
Hirschfelder et al [23]	0,270	6,03
Present correlation	0,279	6,03

4. Recommendation Equation

The helium gas viscosity in the present correlation was predict within the Eq. (16).

$$\eta(T,\rho) = \eta_o(T) + \Delta\eta(T,\rho) \tag{27}$$

At low density region, the Chapman-Enskog solution was used (Eq. (26)). This equation is quite simple and accurate to predict a helium gas viscosity at low pressure. This equation depends on the selected parameters for the collition integral and intermolecular force constants. The collision integral correlation used in the present correlation obtained by Assael et. al. for universal gas [24]as shown in Eq. (28).

$$\begin{aligned}\Omega_\eta(T^*) = \exp(&0.46649 - 0.57015(\ln T^*) + 0.19164(\ln T^*)^2 \\ &- 0.03708(\ln T^*)^3 + 0.00241(\ln T^*)^4)\end{aligned} \tag{28}$$

For high density, the exces viscosity is used by Arp & McCarty correlation as shown in Eq. (29) [25].

$$\Delta\eta(T,\rho) = \rho B(T) + \rho^2 C(T) + \rho^3 D(T) \tag{29}$$

The B(T), C(T) and D(T) are constanta. The constant used in the present correlation is shown in Eq. (30).

$$
\begin{aligned}
B(T) &= -47.525/x + 87.679 - 42.074x + 8.331x^2 - 0.587x^3 \\
C(T) &= 547.305/x - 904.87 + 431.404x - 81.45x^2 + 5.369x^3 \\
D(T) &= -1684.388/x + 3331.086 - 1632.191x + 308.804x^2 - 20.191x^3
\end{aligned}
\tag{30}
$$

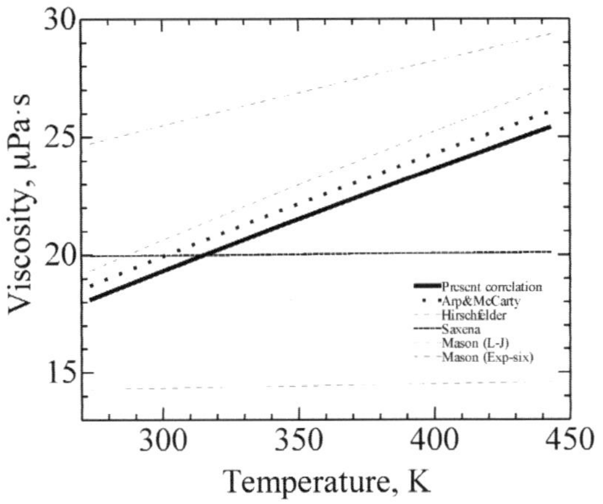

Figure 1. Viscosity prediction at 0.1 MPa based on the intermolecualar force constant used in Table 1. Present correlation and Arp&McCarty correlation give a difference about 2-3 %.

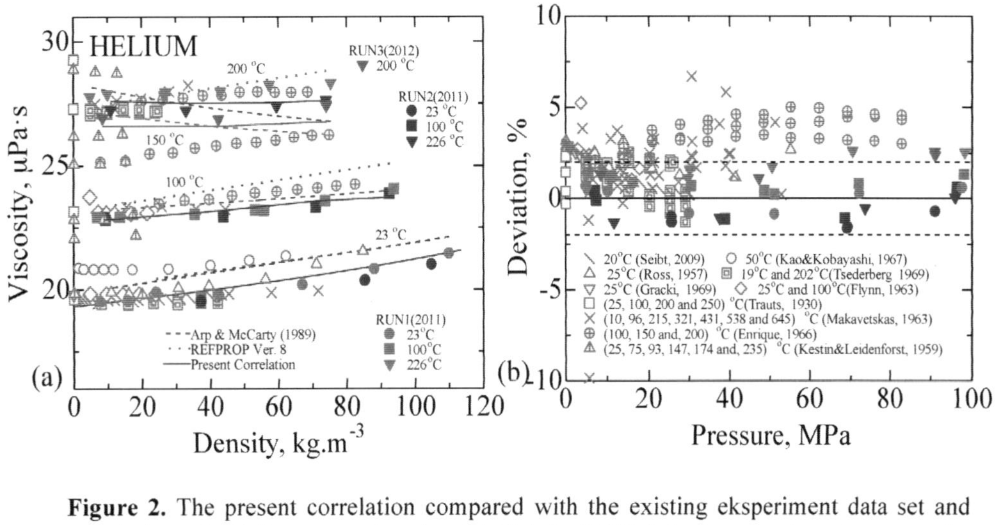

Figure 2. The present correlation compared with the existing eksperiment data set and correlations fo Helium gas viscosity. RUN 1 to RUN 3 is the experiment conduct by HYDROGENIUS [16], The range is 0.1 to 100 MPa and 25 to 226 C.

Figure 1 shows a simulation of helium gas viscosity derived from various intermolecular force constant tabulated in **Table 1**. The discrepancy among them is up to 40%. The present correlation gives a difference about 2-3 % respect to the Arp & McCarty correlation. The tendency of the present works agree well compared with the data obtained by the experiment (HYDROGENIUS) about 2%, and the other researchers up to 5% (Fig. 2). The present correlation is also have been compared with respect to the commersial database developed by NIST (REFPROP)[26]. The deviation is up to ±5 % compared with the present equation. **Figure 3** shows a helium gas viscosity prediction with the present correlation. The prediction has been compared with the experimental data set up from 0,1 to 100 MPa, and 25 to 226 C with the same apparatus described in the Ref. 27 for hydrogen. Based on the prediction shows that for temperatur above 500 K, the trend is opposite for high pressure and low pressure. The present result give a motivation to measure helium at high pressure and temperature in more accurate. The main problem of the measurement so far is about leakage because the viscosity of helium is almost a similar coeficient to hydrogen. The metods to measure helium under extreme condition should be selected very well. The capillary method was not suggesting to measure because of apparatus geometry and limitation on the pressure drop transducer device under high pressure. The suitable method to measure the viscosity is vibrating method, either curve vibrating method or straigh wire.

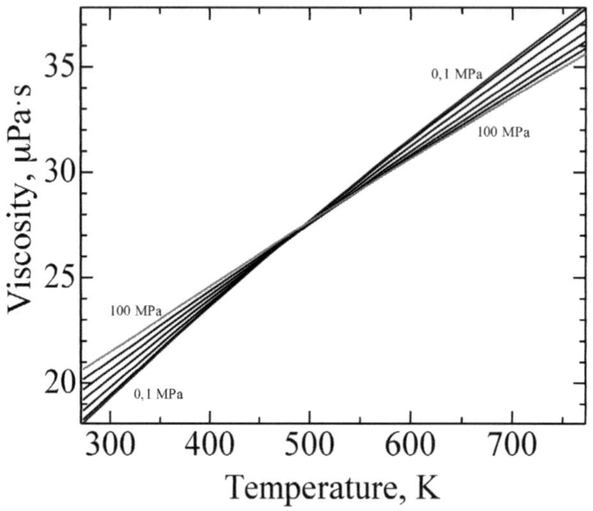

Figure 3. Helium gas viscosity prediction with the present correlation up to 100 MPa, 770 K.

Conclusion

A new correlation has been derived to predict helium gas viscosity at high pressure and temperature. The correlation has been compared with the experimental data set up to 500 K for the temperature and up to 100 MPa for the pressure. The correlation based on the Chapman-Enskog solution and the excess viscosity relation, therefore it is simple and convenient for practical use. The force constants use in the present correlation for the Chapman-Enskog solution are σ=0,279 and ε/κ=6,03, respectively. The exces viscosity in the present correlation use Arp & McCarty's correlation with a few modification. The present correlation gives a difference about 2-3 % respect to the Arp & McCarty correlation.

IOP Conf. Series: Journal of Physics: Conf. Series **799** (2017) 012008 doi:10.1088/1742-6596/799/1/012008

Nomenclature

μ_0	=	Viscosity at 1 atm
T^*	=	Reduced temperature ($=T/(\varepsilon/k)$)
ε	=	A maximum energy of attraction
K	=	the Boltzmann constant
P	=	Density,
Σ	=	Diameter of the molecules
Ω	=	force constants
N	=	Number-density
m	=	Mass of the molecules
X	=	Numerical factor used in Enskoq`s theory related to the collision probability
a	=	Temperature independent function of the molecular forces.
R	=	The gas constant
M	=	The molecular weight.
T	=	Absolute temperature
ψ	=	The collisional transfer of molecular property
C	=	Velocity
P	=	Pressure
p_0	=	Hydrostatic pressure
q	=	Total flux-vector of energy
ρ_0	=	Density of the gas at normal pressure and temperature
c_v	=	Specific heat per unit mass at constant volume
f	=	Maxwell`s velocity distribution function

Reference

[1] F. M. Mitenkov,A. V. Vasyaev,N. G. Kodochigov, V. F. Golovko, N. N. Ponomarev Stepnoi, N. E. Kukharkin and A. Ya. Stolyarevskii, High-Temperature Gas-Cooled Reactors–Energy Source For Industrial Production Of Hydrogen, *Atomic Energy*, Vol. **97**, No. 6, 2004

[2] Dwi Irwanto, Toru Obara, "Decay heat removal without forced cooling on a small simplified PBR with an accumulative fuel loading scheme", *Annals of Nuclear Energy* Vol. 60, pp. 383-395 (2013)

[3] J. O. Hirschfelder, C. F. Curtis And R. B .Bird, *Molecular Theory Of Gases And Liquids*, John Wiley & Sons, 1954

[4] E. Yusibani, P. L.Woodfield, K. Shinzato, M. Kohno, Y. Takata, M. Fujii, Prediction Of Hydrogen Gas Viscosity At High Pressure And High Temperature, *Jpn. J. Thermophys*. 24, 21, 2010

[5] Daniel Seibt, Sebastian Herrmann, Eckhard Vogel, Eckard Bich, And Egon Hassel, Viscometer-Densimeter Over A Wide Range Of Temperature And Pressure, *J. Chem. Eng. Data* 2009, 54, 2626–2637

[6] J. F. Ross and G. M. Brown, *"Viscosity of Gases at High Pressure"*, Ind. Eng. Chem., 49 (1957) 2026

[7] J. A. Gracki, G. P. Flynn, And J. Ross, Viscosity Of Nitrogen, Helium, Hydrogen, And Argon From -100 To 25 C Up To 150 – 250 Atm, , *J. Chem. Phys*.,51(1969)3856

[8] V.M.Trautz And H.E.Binkele, Die Reibung, Warmeleitung Und Diffusion In Gasmischungen VIII. Die Reibung Des H2, He, Ne, Ar Und Ihrer Bindren Gemische, *Annalen Der Physik* 397-5-561 (1930)

[9] R.A.Makavetskas, V.N.Popov And N.V. Tsederberg "Experimental Study Of The Viscosity Of Helium And Nitrogen" *Teplofizika Vysokikh Temperatur* 1(2) 191-7 1963

[10] Enrique G. Reynes And George Thodos, Viscosity Of Helium, Neon, And Nitrogen In

The Dense Gaseous Region, *Journal Of Chemical And Engineering Data* Vol. 11, No. 2, April 1966 Pp 137-140

[11] J.T.F Kao And R. Kobayashi, Viscosity Of Helium And Nitrogen And Their Mixtures At Low Temperatures And Elevated Pressures, *Journal Of Chemical Physics* Vol 47 No 8 1967

[12] N.V. Tsederberg, V.N.Popov And N.A. Morozova, Thermodynamic And Thermophysical Properties Of Helium (Translated From Russian*) Israel Program For Scientific Translations Jerussalem* 1971

[13] J. Kestin Dan W. Leidenfrost, The Viscosity Of Helium, *Physica* 25 537-555 1959

[14] C. Evers, H. W. Losch, And W. Wagner, An Absolute Viscometer-Densimeter And Measurements Of The Viscosity Of Nitrogen, Methane, Helium, Neon, Argon, And Krypton Over A Wide Range Of Density And Temperature, International Journal Of Thermophysics, Vol. 23, No. 6, November 2002 1411-1439

[15] G.P.Flynn, R.V.Hanks, N.A.Lemaire And J.Ross, Viscosity Of Nitrogen, Helium, Neon And Argon From -78.5 To 100 C Below 200 Atmospheres

[16] (HYDROGENIUS team) Elin Yusibani, Yosuke Nagahama, Kosuke Yoshimura, Temujin Uehara, Kan'ei Shinzato, Yasuyuki Takata, Masamichi Kohno, "Measurements Of Hydrogen Viscosity In The Range Of Room Temperature Up To 500 C, And Pressures Up To 100 Mpa With A Capillary Tube Method" *International H2 Energy Development Forum 2012, Workshop In Thermophysical Properties*, Fukuoka 2012

[17] S. Chapman And T.G. Cowling, *The Mathematical Theory Of Non-Uniform Gases*, Cambridge Press, Chp.16. (1958)

[18] D.Enskoq,K. Svensk.*Vet.-Akad.Handl.* 63,No.4 (1921)

[19] A. Michels, A.C.J. Schipper, And W.H. Rintoul, *Physica* (Amsterdam) 19:1011 (1953).

[20] A. Michels, And A. Botzen, *Physica* 18:595 (1953).

[21] Edward A Mason And William E. Rice, The Intermolecular Potentials Of Helium And Hydrogen, *The Journal Of Chemical Physics* Vol. 22 No. 3 1954

[22] S.C.Saxena And B.P. Mathur, Thermaldiffusion In Binary Gas Mixtures And International Forces, *Reviews Of Modern Physics* Vol. 37, No. 2 1965

[23] J.O. Hieschfelder, R. Byron Bird And Ellen L. Spotz, The Transport Properties For Non Polar Gases, *The Journal Of Chemical Physics* Vol 16 No 10 1948

[24] M. J. Assael, S.Mixafendi and W. A. Wakeham, *J.Phys.Chem.Ref.Data, Vol. 15, p.*4 (1986)

[25] Vincent D. Arp and Robert D. McCarty, Thermophysical properties of Helium-4 from 0.8 to 1500 K with Pressure to 2000 MPa, *NIST Technical Note* 1334

[26] E.W. Lemmon, M.L. Huber, M.O. McLinden, NIST Standard Reference Database 23, NIST Reference Fluid Thermodynamic and Transport Properties Database (REFPROP): Version 8 (NIST, Boulder, 2007)

[27] Elin Yusibani, Yosuke Nagahama, Masamichi Kohno, Yasuyuki Takata, Peter L. Woodfield, Kanei Shinzato, Motoo Fujii , A Capillary Tube Viscometer Designed for Measurements of Hydrogen Gas Pressure and High Temperature, *Int. J. Thermophys. Vol.32, pp.1111, 2011*

IOP Conf. Series: Journal of Physics: Conf. Series **799** (2017) 012009 doi:10.1088/1742-6596/799/1/012009

Preliminary Study of Lead-Oxide Cooled Fast Reactor with Natural Uranium as an Input Fuel with Reactor Shuffling Strategy

Rida SN Mahmudah[1*], Zaki Su'ud[2]

[1]Universitas Negeri Yogyakarta, Jl. Colombo No.1, Yogyakarta 55281, Indonesia
[2]Institut Teknologi Bandung, Jl. Ganeca No.10, Bandung 40132, Indonesia

*Email: rida@uny.ac.id

Abstract. A preliminary study of lead-oxide cooled fast reactor with natural uranium as an input fuel using reactor shuffling strategy has been conducted. In this study, reactor core is divided into four zone with the same volume, each zone use different uranium enrichment. The enrichment number is estimated so that in the end of reactor's operation, we only need to add natural uranium as the fresh input fuel. This study used UN-PuN as the fuel and lead oxide as the coolant. Several parameter studies have been conducted to determine the most suitable input condition. It is confirmed in this study that with fuel : cladding : coolant ratio of 53 : 10 : 37, and uranium enrichment in the first to the fourth zone of 0%, 6.25%, 7.5% and 8%, respectively, the reactor can operate as long as 20 years of operation with terminal *k-eff* of 1.0004.

1. Introduction

Nowadays, nuclear energy has become one of the alternatives for energy resources because of its several advantages, such as no greenhouse gas emission, low cost, realiable energy, small volume of waste, etc. However, the enrichment of natural uranium as nuclear power plant (NPP) fuel has been a very sensitive issue since it can be malfunctioned to produce nuclear weapons. This issue has brought IAEA to make Nuclear Non-proliferation Treaty (NPT) in 1968, for policing the nuclear activities of member countries. The previous study has been conducted to offer a solution for the NPP system without the strong dependency on fuel enrichment [1]. The study used Lead-cooled Fast Reactor (LFR) with UN-PuN as the fuel and Pb-Bi as the coolant. The reactor could operate more than 20 years and used natural uranium as an input with reactor shuffling strategy, which will be discussed in the next section. To find a better performance of the reactor, this study use lead-oxide as the coolant instead of lead-bismuth. We start the the preliminary design study by perform a parametric survey and determine the average power density per mesh to get the optimal design of the reactor.

2. Design Concept

In order to utilize natural uranium as the fuel input, we use reactor shuffling strategy. The reactor core is divided into four zone with the same volume, and different enrichment number. In the end of one operation cycle, the fuel in zone 1 (natural uranium) can be moved to zone 2 with the higher enrichment. The fuel in the zone 2 can be moved to the 3rd zone, and so on. This moving-fuel to the next zone is possible by dividing each zone with the same volume. In the end of the operation, we only need to input natural uranium in the 1st zone in the beginning year of the next reactor operation cycle. The diagram of this strategy is shown in Figure 1.

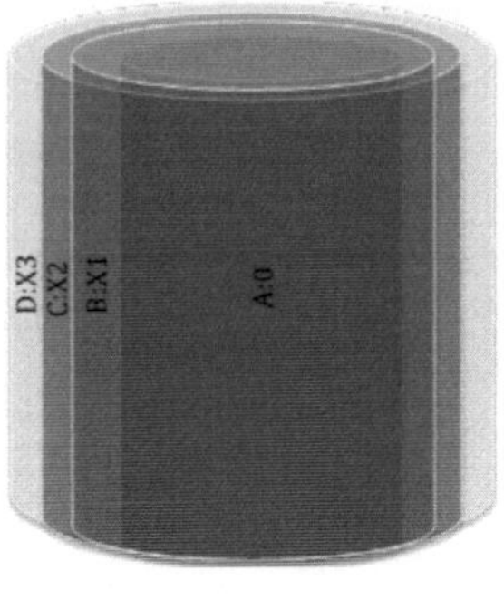

Fig 1.a

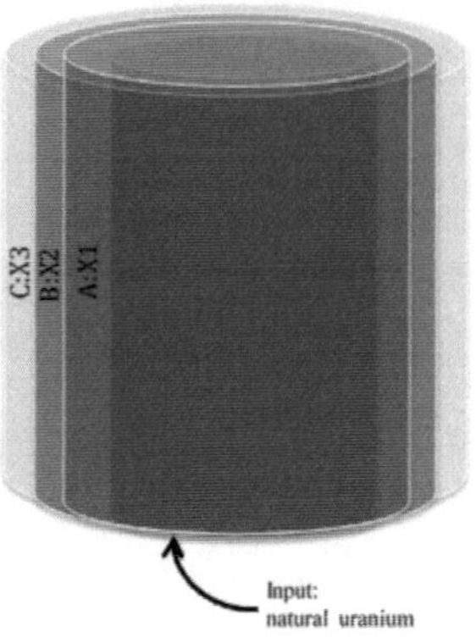

Fig 1.b

After divided the core into four zone, we filled zone A with natural uranium, zone B with plutonium content X1, zone C with plutonium content X2 and zone D with plutonium content X3, with X3 > X2 > X1 (see Figure 1.a). After one operation cycle without refuelling, the plutonium content in zone A will be large enough to be filled in zone B, the plutonium content in zone B will be large enough to be filled in zone C, and the plutonium content in zone C will be large enough to be filled in zone D (see Figure 1.b). Consequently, we only need to input natural uranium in zone A for the next operation cycle.

This study used Lead-cooled Fast Reactor with Pb_3O_4 as the coolant. The specifications of the reactor and the coolant are listed in Tabel 1 and Tabel 2, respectively.

Table 1. Reactor Spesifications

Parameter	Spesification
Power Rate	100 Wt/cc
Refueling periode	More than 20 years
Core Geometry	Cylinder *Balance*
Core Characteristics	*Small-long life core*
Fuel	UN and PuN
Cladding	SS316
Coolant	Pb_3O_4
Pin cell Type	Hexagonal cell
Pin pitch/diameter	1.25 cm
Core height	150 cm
Core diameter	250 cm

Table 2. Pb_3O_4 Physical Properties [3]

Molecular Weight	685.5976 g/mol
Density	8.3 g/cm^3
Melting point	500° C

The study carried out by calculate several parametric studies and determine the average power density per mesh, which will be discussed in the next session.

3. Calculation Method

The calculation in this study is provided by SRAC95 (Standard thermal Reactor Analysis Code system), a program developed by JAERI (Japan Atomic Energy Research Institute). With JENDL-3.2 nuclide data, SRAC will do the calculation and provide microscopis and macroscopic cross-section data from each core's material. The steps are, first, SRAC do the cell and burn-up calculation for every fuel cell, then it being homogenized and collapsed according to determined group. Those calculations will keep occur corresponding to the determined burn-up years. The results of these calculations will be stored in user library. Then, the data from this library will be used in CITATION module to find multiplication factor, reactivity and power density distribution of the core [2]. The flowchart of SRAC calculation is shown in Figure 2.

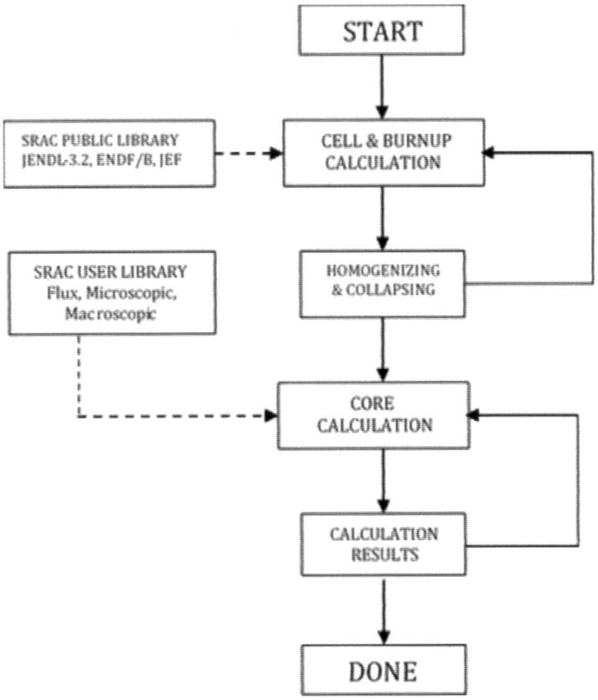

Fig. 2. Flowchart of Reactor Design Calculation by SRAC

IOP Conf. Series: Journal of Physics: Conf. Series **799** (2017) 012009 doi:10.1088/1742-6596/799/1/012009

4. Results and Discussion

4.1. Average Power Density Iteration

From the previous study, we divided the reactor core into four zones: Zone 1 (0% plutonium enrichment), Zone 2 (6.25% plutonium enrichment), Zone 3 (7.5% plutonium enrichment) and Zone 4 (8% plutonium enrichment). First, in cell and burn-up calculation, we input a power density number in for all zones and all burn-up years. After homogenizing and collapsing, the core calculation will results in average power density per zone per year, which the numbers are different from the one that was given at first. The results from this core calculation are then become input for the cell and burn-up calculation for the next iteration, and so on. We keep doing the iterations until we get a reasonably unchanging average power density per zone per year.

In this study, we input 100 W/cc or 1.23×10^{-4} MW/cm power density in the first cell and burn-up calculation, and after 10 iterations, we get average power density in each zone as follow:

Table 3. 10th iteration result for average power density

burn-up year	Average Power Density (MW/cm)			
	Zone 1	Zone 2	Zone 3	Zone 4
1	6.95E-07	1.53E-04	4.71E-04	3.56E-04
2	7.56E-07	1.60E-04	4.73E-04	3.47E-04
3	8.13E-07	1.66E-04	4.74E-04	3.40E-04
4	8.71E-07	1.72E-04	4.75E-04	3.33E-04
5	9.30E-07	1.78E-04	4.75E-04	3.27E-04
6	9.91E-07	1.84E-04	4.74E-04	3.21E-04
7	1.05E-06	1.90E-04	4.74E-04	3.16E-04
8	1.13E-06	1.97E-04	4.72E-04	3.10E-04
9	1.20E-06	2.04E-04	4.71E-04	3.05E-04
10	1.28E-06	2.10E-04	4.69E-04	3.00E-04
11	1.36E-06	2.17E-04	4.67E-04	2.95E-04
12	1.45E-06	2.25E-04	4.65E-04	2.90E-04
13	1.55E-06	2.32E-04	4.62E-04	2.85E-04
14	1.65E-06	2.41E-04	4.59E-04	2.80E-04
15	1.77E-06	2.49E-04	4.55E-04	2.75E-04
16	1.89E-06	2.58E-04	4.51E-04	2.70E-04
17	2.03E-06	2.67E-04	4.47E-04	2.65E-04
18	2.17E-06	2.77E-04	4.42E-04	2.60E-04
19	2.33E-06	2.87E-04	4.37E-04	2.55E-04
20	2.50E-06	2.97E-04	4.32E-04	2.50E-04
21	2.68E-06	3.07E-04	4.27E-04	2.44E-04

The average power density for each iteration in each zone is shown in Figure 3, while the absolute error for each iteration is shown in Figure 4.

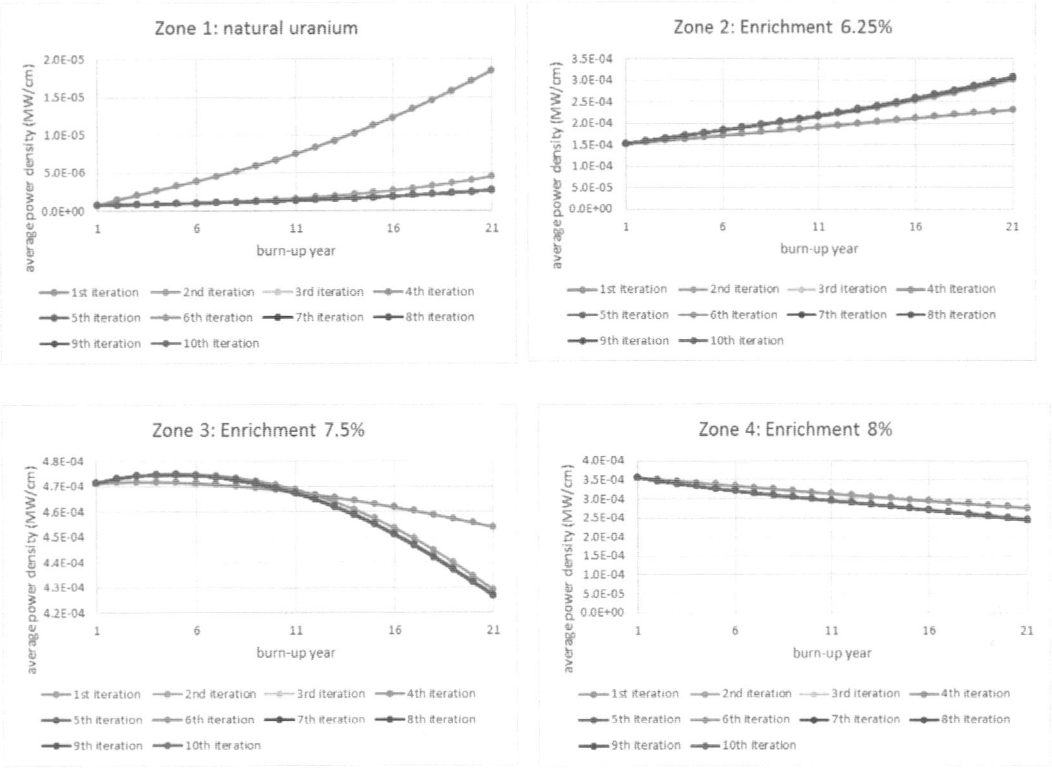

Fig. 3. Iteration results of average power density for each zone

It is shown in Figure 3 that for every zone, the average power density per year results have relatively converge after the 7th iteration. The result from 1st and 2nd iterations shown large difference because we input the same power density for each zone in every burn-up year. As the burn-up and core calculation running, every zones are burned up by different power density due to the differences of numbers of fuel accumulated in each zones in every years of burn-up. Numbers of fuel accumulated in every zones each year can be obtained by doing the core calculation every year. With several iterations, numbers of fuel accumulated in each zones every years are realtively constant. Therefor the average power density also realtively unchanged.

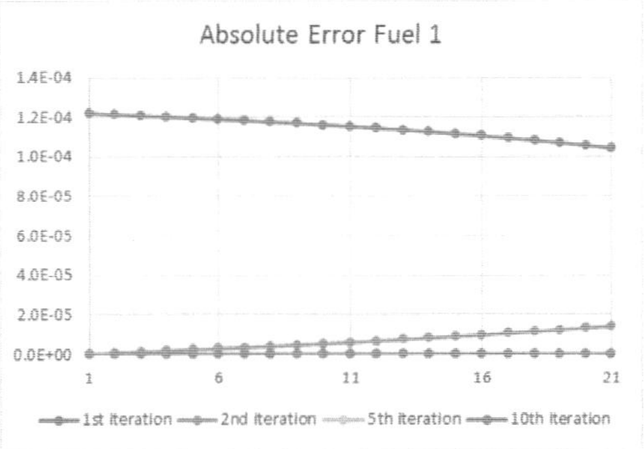

Fig. 4. Absolute Error from several iterations for Fuel 1

In Figure 4, we can see that the absolute error for average power density iteration are reasonably small after the 5th iteration. The absolute error from the 1st to the 2nd iteration is quite large due to the same reason as being describe in power density average iteration above.

4.2. Fuel Fraction Parametric Survey

Several fuel-to-coolant ratio survey calculations have been done in order to get a long life reactor (up to 20 years of operation without refueling). The optimalized fuel-to-coolant ratio has to have k-eff slighty larger than 1 in the 20th year of burn-up, and initial excess relativity less than 10% in the first year of operation. Results from the various fuel-to-coolant ratio is shown in Table 4 below

Table 4. fuel-to-coolant ration survey result

case	fraction			k-eff initial	k-eff final
	fuel	clad	cool		
A	60	10	30	1.1149E+00	1.0296E+00
B	57.5	10	32.5	1.1054E+00	1.0199E+00
C	50	10	40	1.0725E+00	9.8580E-01
D	62.5	10	27.5	1.1239E+00	1.0387E+00
E	55	10	35	1.0951E+00	1.0094E+00
F	54	10	36	1.0909E+00	1.0050E+00
G	53	10	37	1.0865E+00	1.0004E+00

From the table above, we can see that the best fuel:cladding:coolant ratio for this study is 53:10:37. In the first year of its operation, the k-eff is 1.08 and in its 20th, the k-eff 1.0004. The k-eff of every case year by year is shown in Figure 5.

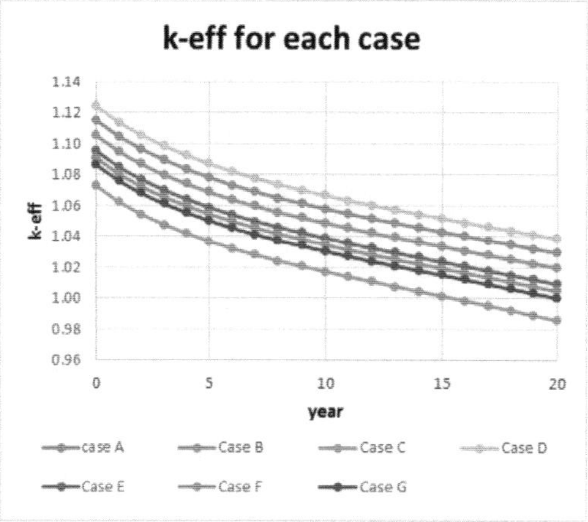

Fig. 5. k-eff for each case

From this fuel-to-coolant ratio survey, we can see that the bigger fuel fraction, the bigger initial and final k-eff. Vice versa, the smaller fuel fraction, the smaller initial dan final k-eff. It because, the bigger fuel fraction, the larger amount of fuel. The larger amount of fuel result in much more fission reactions and much more neutron production. Therefor result in bigger k-eff and vice versa.

5. Conclusion

In this preliminary study, an LFR reactor with UN-PuN as the fuel and lead-oxide as the coolant with natural uranium as the fresh input has been conducted. By utilizing reactor shuffling strategy, the reactor can operate for 20 years without refuelling. The reactor is divided into four zones with different plutonium enrichment in every zone. Several step of iterations has been conducted to determine the appropriate average power density in every zone every year. By doing several fuel-to-coolant ratio, it is confirmed that the reactor can operate for 20 years with fuel:cladding:coolant ratio of 53:10:37.

Acknowledgment

The present study was accomplished with a full support from Nuclear Reactor Physics Laboratory, Department of Physics Bandung Institute of Technology (ITB).

References

[1] Rida SNM (2006) 'Design study of Pb-Bi cooled fast reactors which fuel cycle input is natural uranium' [in Indonesian], BSc Final Project, ITB.
[2] Okumura, K. (2002) 'SRAC: the comprehensive neutronics calculation code system', JAERI Report.
[3] *http://pubchem.ncbi.nlm.nih.gov/compound/22833428*

Monte Carlo Modeling of Gamma Ray Backscattering for Crack Identification in the Aluminum alloy Plate

Rahadi Wirawan[1,], Abdul Waris[2,*], Mitra Djamal[2], Gunawan H.[2], H.J. Kim[3]**

[1]Prodi Fisika, FMIPA, Universitas Mataram, Jl. Majapahit 62 Mataram 83125, Indonesia

[2]Prodi Fisika, FMIPA, Institut Teknologi Bandung, Jl. Ganesa 10 Bandung 40132, Indonesia

[3]Radiation Science Research Institute, Department of Physics, KYUNGPOOK National University, *Daegu 702-701, Republic of Korea*

*Email: rwirawan@yahoo.co.id

Abstract. A Monte Carlo simulation study has been conducted of the Cs-137 gamma ray backscattering in the aluminum alloy plate. This simulation was performed in order to identify the existence of the crack in the aluminum alloy plate, the correlation between the backscattering peak and the crack width. We are able to analyze the absorbed energy distribution in the NaI(Tl) scintillation detector. For the experimental measurement, we are using 5 μCi of a Cs-137 gamma source and 2 in. x 2in. NaI(Tl) scintillation detector with the PMT. The aluminum alloy dimension is about 8 cm x 6 cm x 1 cm. The crack model is represented by the slit with the varying width (1 mm, 2 mm, 4 mm, and 6 mm). The existence of a crack is identified by the decreasing intensity of the gamma backscattering energy peak. These predicted results have a good agreement with the experimental measurement.

1. Introduction

Aluminum alloy is a material that mostly used in industry for aircraft, cycling frames, body panel etc. The presence of defects in the material such as invisible crack, holes or local density anomaly gives an effect of the mechanical strength of the material. Investigation of the defect can be conducted by applying the penetrating radiation. In order to develop the one side measurement access of penetrating radiation to the high density material object, gamma ray backscattering is a potential method to be applied.

The presence of crack in a material will be causing a gap in the material and its influences an element volume which probably involve in the scattering interaction. Therefore, it can affect the detected intensity of gamma photon or energy spectrum distribution. Djamal et al. (2013) have investigated the effect of the hole presence in acrylic block to the energy spectrum distribution [1].

As a preliminary study of investigating the presence of crack based on the gamma ray backscattering method, computer simulation is the promising choice to be done. According to the characteristic of the radiation, Monte Carlo (MC) simulation is a probable approach. MC is useful for modeling a large variety of physical situations, prediction of the experimental results, probe/detector design, and performance of an inspection modeling [2,3].

In this work, we investigate the effect of the crack presence in the back of aluminum alloy plate to the gamma ray energy distribution spectrum of Cs-137 based on the gamma ray backscattering measurement. The preliminary investigation was conducted using the MC GEANT4 simulation toolkit.

2. Theory

Intensity of detecting beam in the crack investigation is a result of direct gamma beams and scattering beams from a plate (Figure 1). Mathematically, it's written as the following equation.

$$I_{\text{detected}} = I_{\text{direct}} + I_{\text{scattering plate}} \quad (1)$$

The scattering interaction of photon beam with an elemental volume of the plate can be described as a result of single scattering or the multiple scattering. According to the single scattering approach (Figure 1), there are three steps process i.e. attenuation process by medium before interaction, the interaction process with an elemental volume, and attenuation of the scattering photon in the movement to the detector.

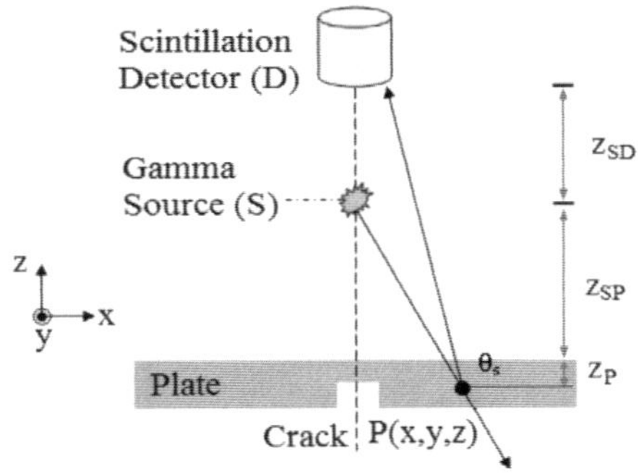

Figure 1. Gamma scattering from the plate model.

Reducing some element volume of the material which described as the presence of crack has an effect on the volume interaction. So, it influences the detected intensity as depicted the detected photon scattering intensity formula in the equation (2),

$$I = \int \left(\frac{S r_0^2 N_A Z A_d}{8\pi M} \right) \rho \times \frac{e^{-\left(\frac{\mu_1(E)}{\rho}\right)\rho r_1}}{r_1^2} \frac{e^{-\left(\frac{\mu_2(E)}{\rho}\right)\rho r_2}}{r_2^2} \times \left(P - P^2 \sin\theta + P^3\right) \sin\beta \cos\phi \, dV \quad (2)$$

and it depends on the source activity (S), classical electron radius (r_0~2.82 fm), Avogadro number (N_A), atomic number (Z), detector surface area (A_d), atomic mass of element volume (M), medium density (ρ), element volume interaction (dV), distance photon source, the vertical angle of element volume position (ϕ) and photon scattering to the element volume interaction (r_1 and r_2) and the scattering parameter (P). The scattering parameter P is depending on the photon energy (E) and the scattering angle θ [5].

$$P(E,\theta) = \frac{1}{1 + \frac{E}{m_0 c^2}(1 - \cos\theta)} \quad (3)$$

where m_0c^2 is an electron energy (0,511 MeV) [4,5].

3. Methodology

In this work, the simulation modeling using GEANT4 is constructed based on an experimental measurement design as shown in Figure 2. The aluminum alloy dimension is about 8 cm x 6 cm x 1 cm. Aluminum alloy material composition for the simulation based on the EDS test result consisting of Mg 1.09%, Al 94.55% and Cu 4.36%). The crack model was represented by a vertical slit with the varying width (1 mm, 2 mm, 4 mm, and 6 mm) and 0.5 cm height. The position is on the bottom surface of the Aluminum alloy plate at the axis line.

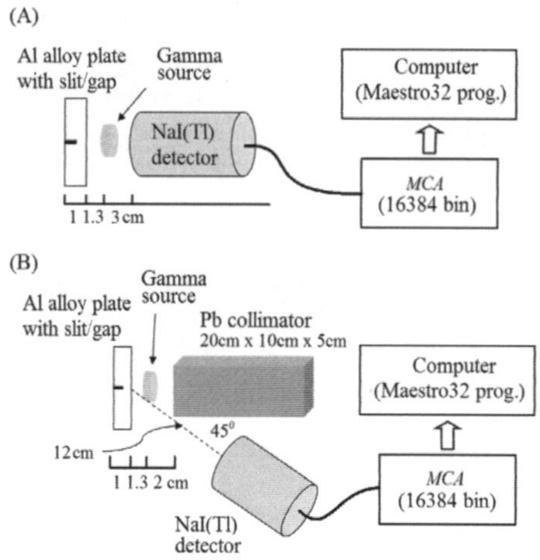

Figure 2. The schematic design of scattering measurement setup.

The number of beamon (history) was used is about $10^7$and a source model of gamma ray in this simulation is a point source. In addition, for the interaction package of MC simulation is *Penelope (Penetration and Energy Loss of Positrons and Electrons)*. Figure 3 shows ones of the GEANT4 visualization of the measurement design which developed in simulation.

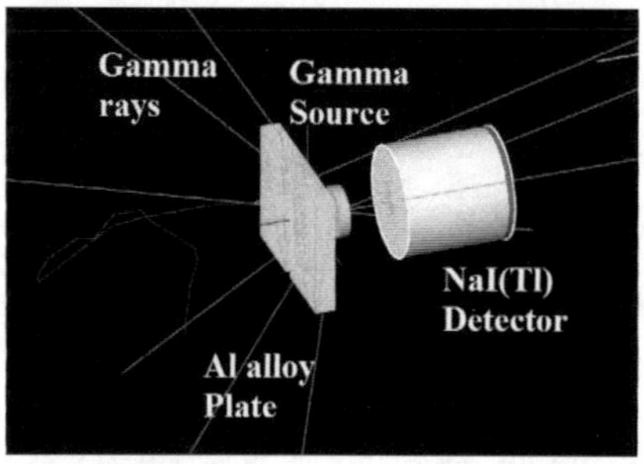

Figure 3. GEANT4 visualization of A type measurement.

For the measurement, a gamma source 5 μCi of Cs-137 and the Bicron 2M2/2 SN IS799 photomultiplier with a NaI(Tl) scintillation detector are used. The high voltage setup of photomultiplier is about 800 V, fine gain and coarse gain amplifications are set in 1.08 and 50 respectively. For data acquisition, MCA with 16384 channel bin is used and it's supported by MAESTRO program.

The optimum counting is chosen based on the recorded spectrum quality. For A type measurement, the counting time is about 5 minutes and 10 minutes for a B type measurement. The recorded energy spectrum will be converted into 400 channel bin using the ROOT program. Therefore, peak curve analysis is conducted to find the peak height of the backscattering energy curve peak.

4. Results and Discussion

The analyze results in the backscattering peak based on Gaussian fitting curve shows that there are reducing the peak height (intensity) in the axis of crack. Figure 4 shows the normalized backscattering (comparison to the backscattering peak of the crack in the central axis).

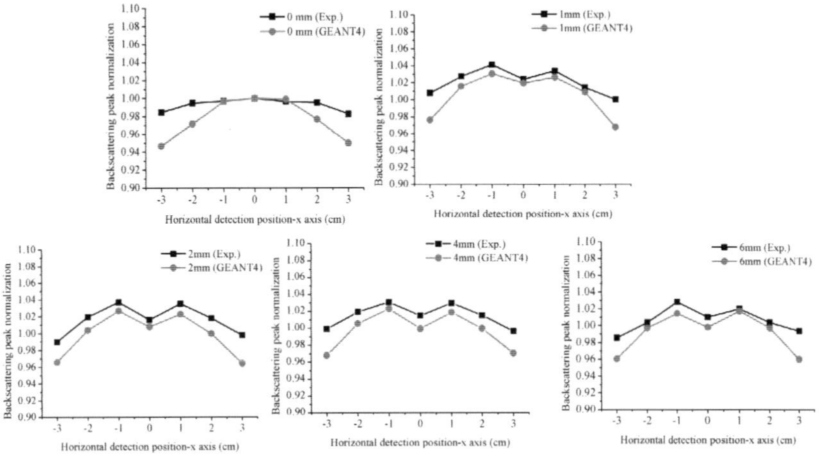

Figure 4. Backscattering peak normalization of A type design.

The similar result is gotten for the B type measurement design as shown in Figure 5. According to these results, the presence of a crack is identified based on lowering backscattering peak height at x=0 detector position compared to the backscattering peak at x = -1 and x=1.

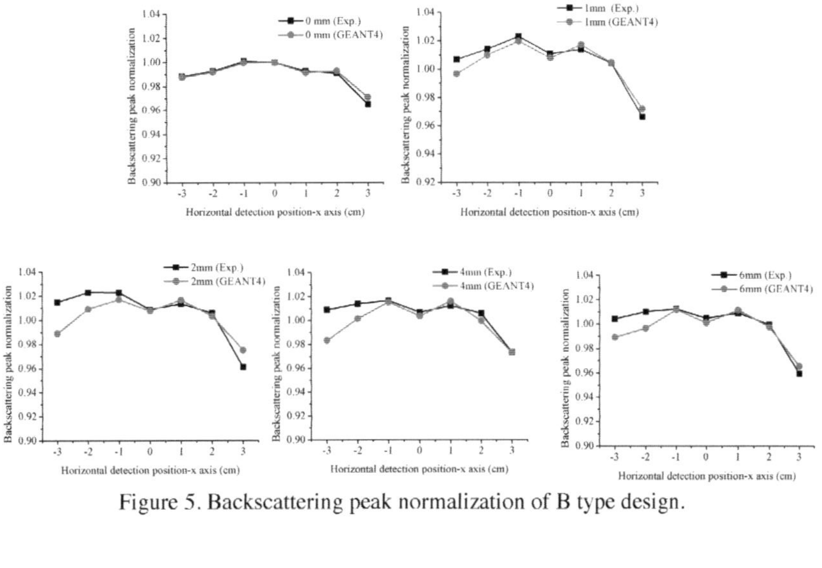

Figure 5. Backscattering peak normalization of B type design.

According to the curve of backscattering peak at x=0 detector position shows that simulation result has the same trend to the measurement as shown in Figure 6. Both simulation and experiment show that an increasing of a crack (slit) width leads the decreasing of backscattering peak height.

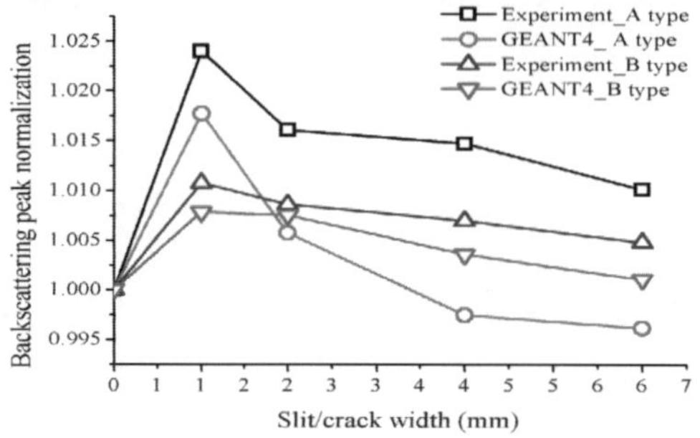

Figure 6. Backscattering peak normalization at x=0 detector position.

Comparing the two measurement type model, MC simulation result for B type gets near to the measurement result than A type. In other hand, the detector counting number for the B type measurement is 7.60 x 10^5 – 8.30 x 10^5 less than A type (2.44 x 10^6 – 2.56 x 10^6). Although the counting time for B type measurement is two times longer than A type measurement. It can explain that the geometry factor of the source – detector arrangement inducing the photon count number. Consequently, this results in an effect to the backscattering peak of the detected energy distribution spectrum.

5. Conclusion

In this study, a Monte Carlo simulation of Cs-137 0.662 MeV backscattering peak was performed in order to conduct preliminary identifying the existence of a crack in the aluminum alloy plate.
The existence of a crack was identified by decreasing intensity of Cs-137 gamma backscattering energy peak. The construction of the source - detector and the crack dimension induce the backscattering energy peak.

References

[1] M. Djamal, R.Wirawan, A. Waris, G. Handayani, and H.J. Kim, "*Determination of an Unknown Volume in the Material based on Gamma Ray Scattering Using GEANT4 Simulation*", JPS Conference Proceedings, The Physical Society of Japan, 2014, 014030, pp.1-4.

[2] P. Calmon, Trends and stakes of NDT simulation, Journal of Nondestructive Evaluation Vol. 31, 339 – 341 (2012).

[3] A.F. Bielajew, *Fundamentals of the Monte Carlo method for neutral and charged particle transport*, Department of Nuclear Engineering and Radiological Sciences, The University of Michigan, 2001, pp.1-8.

[4] A.J. Ball, C.J. Solomon, J.C. Zarnecki, The response of gamma backscatter density gauges to spatial inhomogeneity - An extension of the single scattering model, Nuclear Instruments and Methods in Physics Research B 140 (1998) pp. 449-462.

[5] (5) Knoll, G.F., Radiation Detection and Measurement, second ed., Wiley and Sons, New York, 1989.

Charged Particle Flow Base On Mesoscale Simulation With Coupling MPCD-MD Method In Two Dimension Channel

Annas Nasrudin[1*], Sparisoma Viridi[2] and Dwi Irwanto[2]

[1]Physics Department, Universitas Islam Negeri Sunan Gunung Djati,
Jl. A.H Nasution 105, Bandung 40614, Indonesia
[2]Nuclear Physics and Biophysics Research Division, Institut Teknologi Bandung,
Jl. Ganesha 10, Bandung 40132, Indonesia

*Email: a.nasrudin999@gmail.com

Abstract. Ferrofluid have magnetic behavior. In this review, has study about mesoscale colloidal system. Coupling MPCD-MD method was constructed to build a charged particle flow base on mesoscale simulation in two dimension channel. Various variation of charge and temperature to observe behavior of viscous fluid. Particle interacted with each other and external field in two dimension system. The fluid is flow in same direction in x coordinate. This simulation that was presented here covers the essential flow effect due charge and temperature in the pipe geometry. Dynamic viscosity is rising when charge. And dynamic viscosity is tend to lower value when temperature is rising.

1. Introduction

Ferrofluid are stable colloidal suspension. It consist single-domain magnetic particle coated with a surfactant in a carrier fluid. The particle has magnetic behavior. And the carrier liquid is typically an oil or water base [1].

A ferrofluid can be affected by an external force, such as magnetic field. If magnetic field has applied, it can be logic to a ferrofluid. Can control the flow of the ferrofluids. Benefit application of ferrofluid it is used in various fields, such as mechanical engineering, nuclear reactor, thermal engineering and aerospace [2–5].

In order to investigate, there is lot of design and experimentation of porous media of ferrofluid application, such as numerical experiment of ferrofluid flow phenomena in different geometries in the presence of an external magnetic field [6–15]. Strek and Jopek [15], simulated heat transfer through a ferrofluid under the influence of magnetic dipole.

Ganguly et al. [17] simulated a Two-dimensional pressure-driven flow of a magnetic fluid in a channel to investigate the influence of line source dipole magnetic field on the convective heat transfer. Youcef Amirat [18], research a differential system introduced by M.I. Shliomis to describe the motion of a ferrofluid driven by an external magnetic field. That combination of the Navier-Stokes equation.

Habib Aminfar [19], simulated the interactions of magnetic nanoparticles are based on magnetic dipole–dipole interaction potential. Simulations reveal that the system with magnetic nanoparticles agglomerates faster than that of non-magnetic nanoparticles.

P. K. Papadopoulos [20], investigate the accuracy of an analytical solution for the magnetization equation and assess its validity when used for non-uniform magnetic fields. It is found that it can be very helpful as a means of estimating the magnetization, especially for strong magnetic fields with low gradients; second, to examine the effects of the magnetic field on the flow and study the relevant importance of the magnetic terms of the momentum equation. It is revealed that the axial pressure drop depends linearly on the volumetric concentration and that the magneto viscosity effect is negligible in cases of non-uniform magnetic fields.

Many investigation of ferrofluids behavior, which are study the hydrodynamic of flow single phase model of colloid mixture and polymer-colloid mixture. Inside MPCD algorithm is a Stochastic Rotation dynamic to ensure that collision change the direction and the velocity. But MPCD algorithm come to summaries of cell average. Which mean the particle is only interacted with cell.

For solving when a single particle is experience a force with other particle or experience an interference from the outside system then we add MD method. So in this time we apply a coupling MPCD-MD in this numerical simulation. This MPCD method is adapted from [21]. In the MPCD algorithm we use to find momentum and angular momentum, however SRD unable to calculate the angular momentum. Charge fluids show that many behavior and rheological characteristics when it flowed [22].

Takehiro Yamamoto [23], create a numerical simulation of star polymer under shear flow using a coupling method of multi-particle collision dynamics and molecular dynamics. The numerical results agree well with both experimental results and predictions by other numerical methods.

2. Numerical Methods

Numerical methods in this paper we modeled a charge fluids flow under Cartesian coordinate. This fluid were represented by particle. In numerical case, MD and BD simulation was performed and studied by many researcher. Larson and Edwards [24], simulation the motion of charged fluids was modeled by MD, and the flow of fluid computed via MPCD. MPCD algorithm and MD algorithm was coupled to consider interaction of charge particle and other external field. Then the hydrodynamic interaction also considered with MPCD.

The Cartesian coordinate system was employed. The direction of flow was and the velocity profile was . the system length is and height is , for MD and MPCD computation. The system is divided in to a small cell with length a, which is declared to be collision cells.

Assume we have system with particle then distributed randomly into 2D channel of volume . And particle has position and velocity , . MD is part of solving newton's Equation of motion which many particle in cell interacting with each other.

$$\vec{F}_i = m_i \frac{d^2\vec{r}_i}{dt^2} = \partial \frac{\vec{U}(r_1, r_2, \cdots, r_n)}{\partial r_i} \qquad (1)$$

Where m_i is particle mass of i-th particle, r_i the particle position. Using velocity verlet method that can give the position, velocities and acceleration of particle in the system as a function of a time. In this study, we represent of ferrofluid flow as a system of a charged particles.

On the MD method, the equation for liquid particle inside a system is:

$$m_i \frac{d^2 r_{ij}}{dt^2} = \sum_{j \neq i,\, j=1}^{N} \vec{F}_{ij} + \sum_{j_w \neq i,\, j=1}^{N_w} \vec{F}_{ij} + \vec{F}_{ext} \qquad (2)$$

Where i, j represent of particle, and $\vec{i}$ is the unit vector in xy-coordinate. The first section of the equation is the potential force between particle i and all particle j in the cell. The last term is represents the external force, such as magnetic field and electric field.

$$\vec{F}_{ij} = k\frac{q_i q_j}{r_{ij}}\frac{\vec{r}_{ij}}{r_{ij}} \qquad (3)$$

And then to calculate potential force between particles in the cell, using coulomb equation (3) to find total force on each particle in the collision cell. That denoted by

$$\vec{F}_i = \sum_{i\neq j}^{N} \vec{\mathrm{F}}_{\mathrm{ij}} \qquad (4)$$

Using velocity verlet method particle i interacted in the small cell of system to reduce amount of the total interacted particle in the system. Because the part of simulation who take lot of time is updating the velocity. Thereby array separation is made from particle data, is changed in to link cell binning [25].

In the case of external field we use the magnetic field. The direction the magnetic field is a cross product, the magnetic force on a charge q, moving with velocity in a magnetic field B is

$$\vec{F}_{mag} = q(v \times B) \qquad (5)$$

This is known as Lorentz force law, in the presence of both electric and magnetic fields, so total force in q would be

$$\vec{F}_{ext} = q[E + (v \times B)] \qquad (6)$$

At this simulation, the magnetic field is coming from solenoid that magnetic field has given value by B.

So equation (4) by newton 2 law of motion, become

$$\sum \vec{F} = m_i \vec{a}_i \qquad (7)$$

That $\sum F$ is come from equation 3 and 6. Thereby

$$\vec{F}_i + \vec{F}_{ext} = m_i \vec{a}_i$$

$$\frac{\vec{F}_i + \vec{F}_{ext}}{m_i} = \vec{a}_i \qquad (8)$$

That force change the acceleration, which correlated to each particle velocity.

3. Simulation

The fluids is represented by same charged particle with mesoscopic scale.

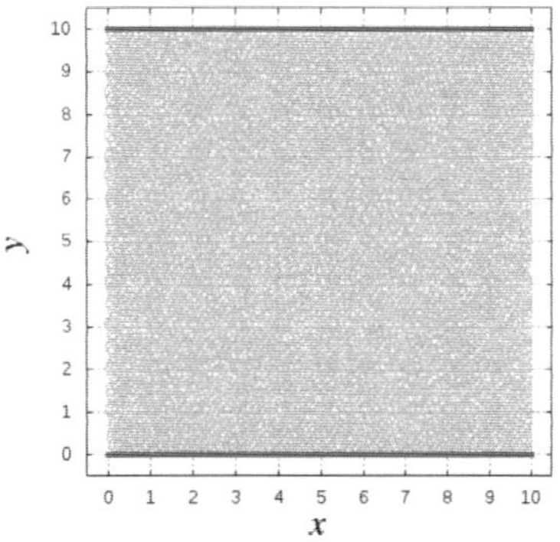

Figure. 1. Fluid in xy-coordinate, which is represented by particle.

First, the particles is interacted with other particle in cell, which is can experience a coulomb interaction with fellow. When there is an external forces, the particles interacted simultaneously with that external forces. So in this part, moving particle is calculate by Molecular Dynamic.

The motion of charge particle was computed using velocity verlet method expressed by following equation:

$$\vec{r}_i(t+\Delta t_{MD}) = \vec{r}_i(t) + t_{MD}\vec{v}_i(t) + \frac{(t_{MD})^2}{2m_i}\vec{F}_i(t) \quad (9)$$

$$\vec{v}_i(t+\Delta t_{MD}) = \vec{v}_i(t) + \frac{\Delta t_{MD}}{2m_i}\left[\vec{F}_i(t+\Delta t_{MD}) + \vec{F}_i(t)\right]$$

(10)

Where r_i , v_i and m_i are the position vector the velocity vector, and the mass of particle. Then Δt_{MD} show that the time increment in the MD simulation part. So

$$\Delta t_{MD} = \Delta t_{MPCD}/5 \quad (11)$$

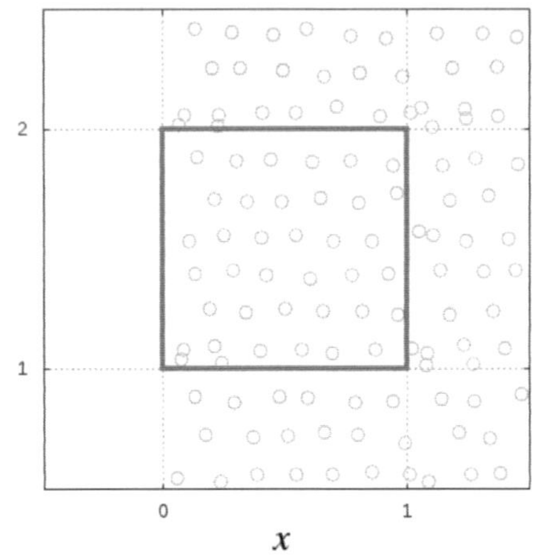

Figure. 2. Small part of system (cell), this cell refer to collision cell. The cell is divided by small red rectangle. Any particle in this cell is collide and rotate.

Then this particles is experienced collision with fellow. Not in pair, but collision occurs in a cell at once. That illustrated in fig. 2 who then experience a rotate in SRD.

In MPCD algorithm who occur SRD rotation, come into two step. The streaming step and collision step. In streaming step, particle position $\vec{r}_i(t)$ is updating base on :

$$\vec{r}_i(t+\Delta t_{MPCD}) = \vec{r}_i(t) + \vec{v}_i(t)\Delta t_{MPCD} \quad (12)$$

Which Δt_{MPCD} is time increment in MPCD algorithm. In the second part particle velocity i can be changed by a stochastic rotation matrix R_c, with angle rotation α via

$$v_i(t+\Delta t_{MPCD}) = \vec{u}_c(t) + \mathrm{R}_c\left[\vec{v}_i - \vec{u}_c(t)\right] \quad (13)$$

Where $\vec{u}_c$ is average velocity in cell. In the MPCD collision step show that momentum particle in the cell is redescribe.

Then we apply boundary condition in the system, invers collision in $y = 0$ and $y = H$. And continuously in $x = 0$ and $x = L$.

For computing value of dynamic viscosity and thermal conductivity is come from [26]

$$\eta = \rho k_B T \frac{1 - e^{-\rho}}{2(e^{-\rho} - 1 + \rho)} \quad (14)$$

And

$$\lambda = \rho k_B T \frac{\rho - 1 + e^{-\rho}}{1 + \rho - e^{-\rho}} \quad (15)$$

4. Results and Discussion

With consideration fluid wich have charge in it, fluid representing with particle with same charge. First, we simulating with numerical for interaction charge particle who has given into fluid. At this time we investigating fluid behavior through different viscosity of fluids.

With using parameter $A_0 = 1$, $R_{SHIFT} = 0.5$, $GhostP = 0$, $N_X = 30$, $N_Y = 30$, $N_{AV} = 10$, $m = 1$, $k_B = 1$, $T = 0.4$, $Seed = 1$, $\mathrm{Re}s = 1e - 4$, $Gamma = 0$, $p_I = 20$, $p_O = 10$, $k = 20$, $B = 15$.

Particle interaction with each other and external force was computed by MD algorithm. Any collidion and rotation particle computed by MPCD algorithm. In this section using iteration by $t_{MPCD} = 50$, and $\Delta t_{MD} = 0.2$. We use variation charge each step is 300 coulomb differential.

This fluid was simulated in the same direction with x. And then we analyze using regression analisys. Thereby we calculate the average $f(x)$ to find dynamic viscosity in the system.

The first behavior which we investigate is behavior fluid when we give different fluid temperature at different simulation.

Fig. 3 and 4 has show that relation dynamic viscosity with temperature. Viscosity value is tended to lower value if we observing in a whole simulation. In this simulation when temperature is the free variable there are critical point. In fig. 3 has 1, and in fig 4. has 2 critical point. In this simulation, this critical point we suppose a statistically error due unlinearity graph.

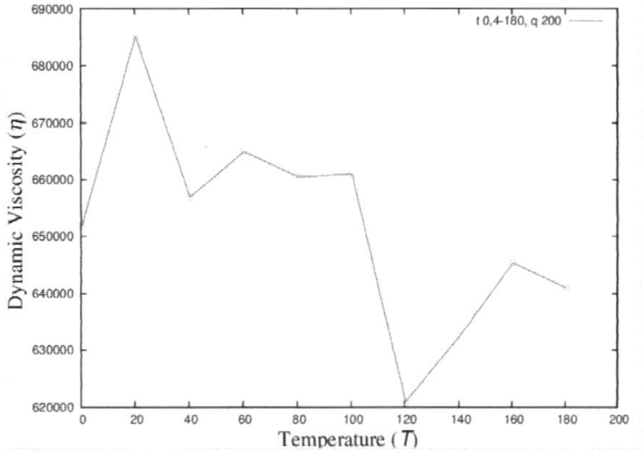

Figure. 3. Comparison of dynamic viscosity with charge function with MPCD-MD method with q = 200

IOP Conf. Series: Journal of Physics: Conf. Series **799** (2017) 012011 doi:10.1088/1742-6596/799/1/012011

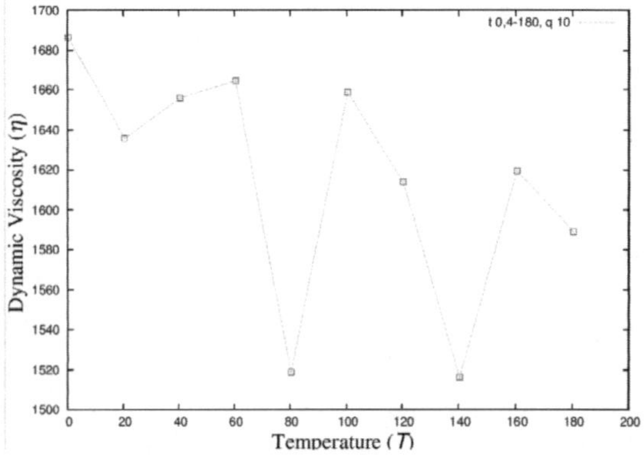

Figure. 4. Comparison of dynamic viscosity with charge function with MPCD-MD method with q = 10

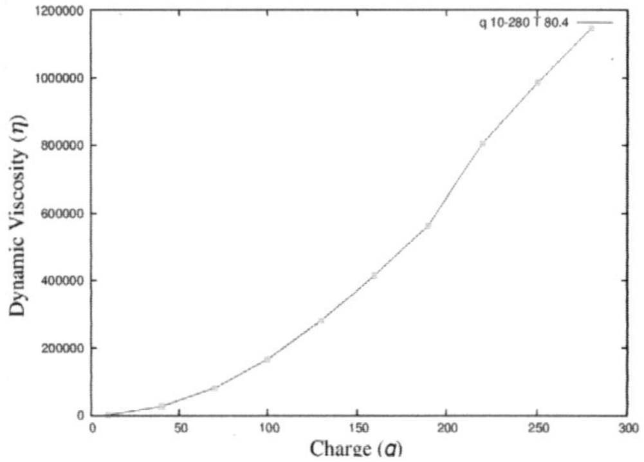

Figure. 5. Comparison of dynamic viscosity with charge function with MPCD-MD method at T = 0.4

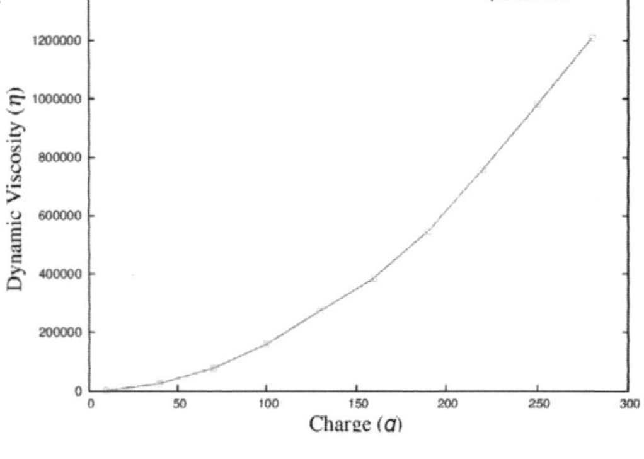

Figure. 6. Comparison of dynamic viscosity with charge function with MPCD-MD method at T = 80.4

Then we investigate another behavior, which is relation between charge fluid and dynamic viscosity. Using parameter $A_0 = 1$, $R_{SHIFT} = 0.5$, $GhostP = 0$, $N_X = 30$, $N_Y = 30$, $N_{AV} = 10$, $m = 1$, $k_B = 1$, $q_p = 10$, $Seed = 1$, $Res = 1e - 4$, $Gamma = 0$, $p_I = 20$, $p_O = 10$, $k = 20$, $B = 15$.

Which $T_{MIN} = 0.4$ and $T_{MAX} = 80.4$. In each step same with step in previously declared which is use 50 time step in each repetition.

Fig. 5 and 6 has show that relation dynamic viscosity with charge. In this simulation we simulating fluid with charge variation. And we test with other different temperature. In this section has show that fluid viscous is rising when we add more charge in to system. This relation is show the linearity between viscosity with charge who has given in the system.

5. Conclusion

Particle base fluid flow simulation at mesoscopic scale was simulated with MPCD-MD method for charged particle. From the result was shown that dependency of viscosity with temperature is tend to lower value. In this case this fluid who have been simulated is fit to generally fluid. Another numerical result and experimental data must be proven to support this investigation. And another case is dependency fluid between viscosity and charge show that linearity relation.

References

[1] Thomas A. Franklin. "Ferrofluid Flow Phenomena", Thesis, June 2003
[2] R.E. Rosensweig, Ferrohydrodynamics, Cambridge University Press, London, 1985.
[3] R. Hiegeister, W. Andra, N. Buske, R. Hergt, I. Hilger, U. Richter, W. Kaiser, Application of magnetite ferrofluids for hyperthermia, Journal of Magnetism and Magnetic Materials 201 (1999) 420–422.
[4] K. Nakatsuka, B. Jeyadevan, S. Neveu, H. Koganezawa, The magnetic fluid for heat transfer applications, Journal of Magnetism and Magnetic Materials 252 (2002) 360–362.
[5] S. Shuchi, K. Sakatani, H. Yamaguchi, An application of a binary mixture of magnetic fluid for heat transport devices, Journal of Magnetism and Magnetic Materials 289 (2005) 257–259.
[6] (6. H. Kikura, T. Sawada, T. Tanahashi, Natural convection of a magnetic fluid in a cubic enclosure, Journal of Magnetism and Magnetic Materials 122 (1993) 315–318.
[7] H. Yamaguchi, I. Kobori, Y. Uehata, K. Shimada, Natural convection of magnetic fluid in a rectangular box, Journal of Magnetism and Magnetic Materials 201 (1999) 264–267.
[8] C. Tangthieng, B.A. Finlayson, J. Maulbetsch, T. Cader, Finite element model of magnetoconvection of a ferrofluid, Journal of Magnetism and Magnetic Materials 201 (1999) 252–255.
[9] C.Y. Wen, C.Y. Chen, S.F. Yang, Flow visualization of natural convection of magnetic fluid in a rectangular Hele–Shaw cell, Journal of Magnetism and Magnetic Materials 252C (2002) 206–208.
[10] M.S. Krakov, I.V. Nikiforov, To the influence of uniform magnetic field on thermomagnetic convection in square cavity, Journal of Magnetism and Magnetic Materials 252 (2002) 209–211.
[11] H. Yamaguchi, Z. Zhang, S. Shuchi, K. Shimada, Heat transfer characteristics of magnetic fluid in a partitioned rectangular box, Journal of Magnetism and Magnetic Materials 252 (2002) 203–205.
[12] S.M. Snyder, T. Cader, B.A. Finlayson, Finite element model of magnetoconvection of a ferrofluid, Journal of Magnetism and Magnetic Materials 262 (2003) 269–279.

[13] R. Ganguly, S. Sen, I.K. Puri, Thermomagnetic convection in a square enclosure using a line-dipole, Physics of Fluids 16 (2004) 2228–2236.
[14] C.Y. Wen, W.P. Su, Natural convection of magnetic fluid in a rectangular Hele–Shaw cell, Journal of Magnetism and Magnetic Materials 289 (2005) 209–302.
[15] T. Tagawa, A. Ujihara, H. Ozoe, Average heat transfer rates measured in two different temperature ranges for magnetic convection of horizontal water layer heated from below, International Journal of Heat and Mass Transfer 49 (2006) 3555–3560
[16] T. Strek, H. Jopek, Computer simulation of heat transfer through a ferrofluid, Journal of Solid State Physics 244 (2006) 1027–1037
[17] R. Ganguly, S. Sen, I.K. Puri, Heat transfer augmentation using a magnetic fluid under the influence of a line dipole, Journal of Magnetism and Magnetic Materials 271 (2004) 63–73
[18] Youcef Amirat, Kamel Hamdach, Strong solutions to the equations of a ferrofluid flow model, J. Math. Anal. Appl. 353 (2009) 271–294
[19] H. Aminfar, M. Mohammadpourfard, S. Ahangar Zonouzi, Numerical study of the ferrofluid flow and heat transfer through a rectangular duct in the presence of a non-uniform transverse magnetic field, Journal of Magnetism and Magnetic Materials 327 (2013) 31–42
[20] P. K. Papadopoulos, P. Vafeas, and P. M. Hatzikonstantinou, Ferrofluid pipe flow under the influence of the magnetic field of a cylindrical coil, Physics Of Fluids 24, 122002 (2012)
[21] A. Nasrudin, S. Viridi, Y. S. Perkasa, Analisis suhu pada simulasi aliran hagen-poiseulli dalam saluran persegi panjang dua dimensi dengan metode multi-particle collision dynamics, Proceeding SNIPS 2015, ISBN 978-602-19655-8-0
[22] Vlassopoulos D, Fytas G, Pakula T, Roovers J (2001) Multiarm star polymers dynamics. J Phys Condens Matter 13:R855–R876
[23] Takehiro Yamamoto, Norichika Masaoka, Numerical simulation of star polymers under shear flow using a coupling method of multi-particle collision dynamics and molecular dynamics, Rheol Acta (2015) 54:139–147
[24] Larson RG, Edwards SF (2005), the rheology of dilute solutions of flexible polymers progress and problems. J Rheol 49:1–70
[25] G. Govindu, R. Scrofano, and V. K. Prasanna. A library of parameterizable floating-point cores for FPGAs and their application to scientific computing. In Proceedings of the International Conference on Engineering Reconfigurable Systems and Algorithms, June 2005.
[26] Anatoly Malevanets and Raymond Kapral,
[27] "Mesoscopic model for solvent dynamics", J. Chem. Phys., Vol. 110, No. 17, 1 May 199

Particle-based Simulation of Rayleigh-Taylor Instability

Nur Asiah Aprianti[1*], Sparisoma Viridi[1], Zaki Su'ud[1], Koji Morita[2]

[1]Nuclear Physics and Bio Physics Research Group, Bandung Institute of Technology, Gedung Fisika, Jl. Ganesha 10, Bandung 40132, Indonesia.
[2]Department of Applied Quantum Physics and Nuclear Engineering, Kyushu University, Moto-oka 774, Fukuoka, Japan

Email: nura@fi.itb.ac.id

Abstract. Rayleigh-Taylor instability is a typical benchmark in a computational fluid dynamics, which show interpenetrate materials happened when a heavy liquid is above a lighter liquid. In particle method, Shirakawa et. al. had been proposed a correction term on modelling a buoyancy force to show the phenomena of Rayleigh-Taylor instability (RTI) in the simulation. By adjust a specific value of correction term for a specific mass density ratio, Rayleigh-Taylor instability phenomena had been observed on MPS method. Later, the phenomena is also observed on FVP methods using the same modelling. In this study, several mass density ratio and viscosity ratio of liquids have been proposed as parametric simulation to observe Rayleigh-Taylor instability. The results show the correlations between mass density ratio and viscosity ratio in the specific correction term value of buoyancy model. This results confirmed that the RTI phenomena can be observed in specific mass density ratio.

1. Introduction

The Rayleigh Taylor Instability is one of fingering instabilities of an interface between two fluids of different densities. It occurs, in many natural and engineering fields, when the light fluids accelerates the heavy fluid. In, computational fluid dynamics, the Rayleigh Taylor Instabilities is a typical benchmark to simulate the buoyancy force.

There are a wide variety of particle-based simulation method; Smoothed Particle Hydrodynamics (SPH), Moving Particle Semi-implicit (MPS), and Finite Volume Particle (FVP). These particle-based simulation methods are powerful on simulate particular shape and geometry since no mesh needed on modelling. In this particle method, we use deterministic approach to solve governing equations. The continuum is represented by particle, where the convection and deformation are simulated by particle's movement.

In this study, the influence of density ratio and viscosity ratio to the Rayleigh Taylor Instability (RTI) phenomena were analysed using particle based simulation.

2. Finite Volume Particle method

In this study, the Rayleigh-Taylor Instability is analysed using a fully Lagrangian approach, based on the FVP method, for incompressible flows. In FVP methods, particle variables such as velocity and pressure are obtained by solving governing equation; the Navier-Stokes equation and the continuity equation.

$$\rho \frac{D\vec{u}}{Dt} = -\nabla p + \nabla(\mu \nabla \cdot \vec{u}) + \vec{f}_s + \vec{f}_g \tag{1}$$

$$\nabla \cdot \vec{u} = 0 \tag{2}$$

where ρ is the density, $\vec{u}$ is the velocity, t is the time, p is the pressure, μ is the dynamic viscosity. The third term $\vec{f}$ in the right-hand side of Eq. (1) represents other forces per unit volume, such as surface tension and gravity.

In the FVP method, the numerical particles are assumed to occupy a certain volume, where the control volume of one moving particle is a circle in 2D simulations.

$$S = 2\pi R, V = \pi R^2 = (\Delta l)^2 \tag{3}$$

where S, V, R, and Δl are the particle surface area, the particle control volume, the radius of particle control volume and the initial particle distance, respectively. According to Gauss's law, the gradient and Laplacian operators acting on arbitrary scalar function ϕ in the FVP method can be approximated as

$$\langle \nabla \phi \rangle_i = \langle \frac{1}{V} \oint_S \phi \vec{n} dS \rangle_i = \frac{1}{V} \sum_{j \neq i} \phi_{sur} \cdot \vec{n}_{ij} \cdot \Delta S_{ij} \tag{4}$$

$$\langle \nabla^2 \phi \rangle_i = \langle \frac{1}{V} \oint_S \nabla \phi \cdot \vec{n} dS \rangle_i = \frac{1}{V} \sum_{j \neq i} \left(\frac{\phi_j - \phi_i}{|\vec{r}_{ij}|} \right) \cdot \Delta S_{ij} \tag{5}$$

In the present study, the natural convection flows, which is driven by the buoyancy force, was considered by the Boussinesq approximation for the gravity term involved in f of Eq. (1).

$$\vec{f}_{g,i} = \rho_i [1 - \beta_i (T_i - T_0)] \vec{g} - \sum_{i \neq j} c_B (\rho_j - \rho_i) \vec{g} w_{ij} \tag{6}$$

where $\vec{g}$, β, T_0, and c_B are the gravity, the volumetric thermal expansion coefficient, the reference temperature, and coefficient correction, respectively.

3. Simulation conditions

Figure 1 shows the geometrical simulation of Rayleigh-Taylor Instability using FVP method. We performed 2D calculations by assumed adiabatic geometrical boundaries, neglecting the effect of heat transfer from the system to surroundings. The initial particle distance is setted at 1.0 mm. The simulation were conducted for 1.5 s with the time step size 1.0 x 10^{-4} s.

IOP Conf. Series: Journal of Physics: Conf. Series **799** (2017) 012012 doi:10.1088/1742-6596/799/1/012012

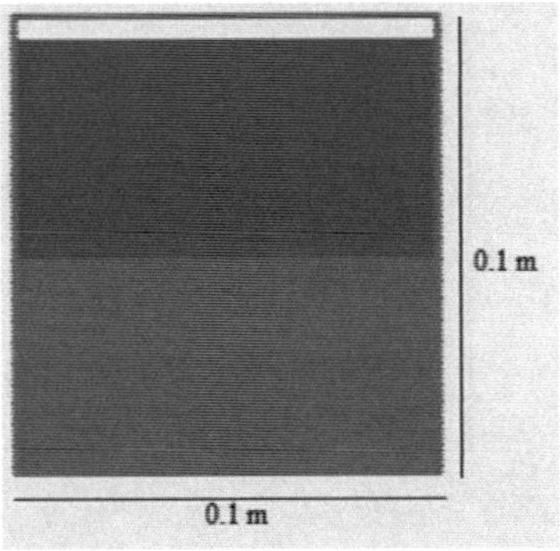

Figure 1. Simulation geometry

To estimate essential effect of density and viscosity ratio to the Rayleigh Taylor Instability (RTI) phenomena, in the present study, we varies mass density ratio and viscosity ratio as shown in table 1.

Table 1. Parametric simulations

	Mass density (kg/m^3)	Surface tension (N/m)	Viscosity (m^2/s)
A	0.90×10^3 - 0.98×10^3	0.0527	1.0×10^{-6} - 96.0×10^{-6}
B	1.0×10^3	0.0728	1.0×10^{-6}

4. Results and Discussion

Figure 2 shows conditions where the Rayleigh-Taylor instability could be observed. There is no effect varying viscosity ratio to the Rayleigh-Taylor instability. Only density ratio which has effect to the phenomena. The Rayleigh-Taylor Instability (RTI) can be observed clearly in specific density ratio 0.91 – 0.97.

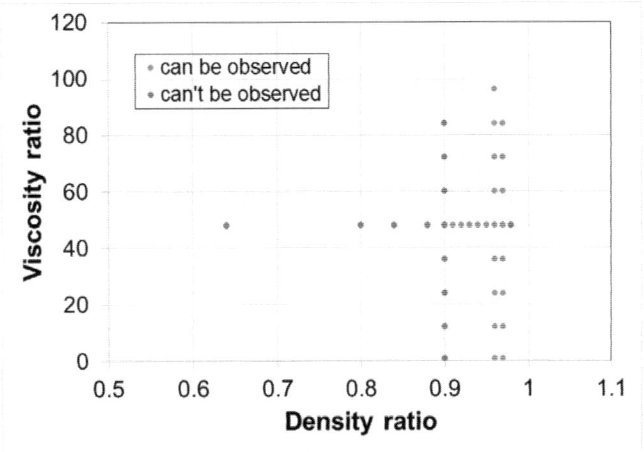

Figure 2. Simulation result (1)

The effect of density ratio can be seen in Table 2. For viscosity ratio 60, the Rayleigh-Taylor instability occurs clearly at density ratio 0.97. Smaller density ratio shows smaller amplitude of fingering instability. In other hand, the Rayleigh-Taylor phenomena could not be observed for higher density ratio (0.98).

Table 2. Effect of density ratio

Viscosity ratio	Density ratio		
	0.98	0.97	0.91
60			

Figure 3. Show the correlation of peak amplitude time and density ratio. This correlation can be approximated by polynomial regression.

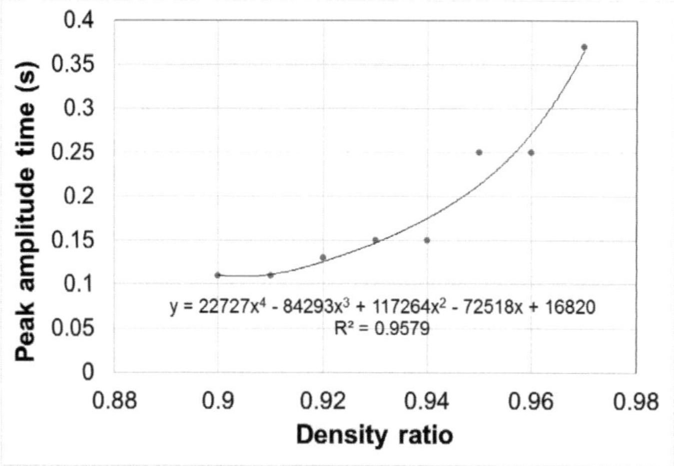

Figure 3. Simulation result (2)

5. Conclusions

The Rayleigh-Taylor Instability (RTI) phenomena can be observed clearly in specific density ratio 0.91-0.97. In smaller density ratio (than 0.91), the buoyancy force is too large so the RTI almost cannot be observed. In larger density ratio (than 0.98), the buoyancy force is too small so the RTI also cannot be observed.

Acknowledgment

Socialication of this research is supported by research program ITB, batch II with contract number 1763/I1.B04.1/KU/2015

References

[1] Rayleigh Lord 1883 Investigation of the character of the equilibrium of an incompressible heavy fluid of variable density *Proceedings of the London Mathematical Society* 14:170-177

[2] Taylor G I 1950 The instability of liquid surfaces when accelerated in a direction perpendicular to their planes *Proceedings of the Royal Society of London* A201:192-196

[3] S Zhang K Morita N Shirakawa dan K Fukuda 2004 Simulation of the Rayleigh-Taylor Instability with the MPS Method *Memoirs of the faculty of Engineering* Kyushu University Jepang Desember Vol 64 No 4

[4] Nur Asiah Aprianti 2014 *Study on Particle-based Simulations of Multiphase Flows with Heat and Mass Transfer for Reactor Safety Analysis* Disertasi Doktor Kyushu University Jepang

[5] Nur Asiah Aprianti 2015 The study of Fluid on Rayleigh-Taylor Instability Phenomena using Finite Volume Particle (FVP) method *Indonesia Journal of Physics* Indonesia Vol. 26 No. 2

IOP Conf. Series: Journal of Physics: Conf. Series **799** (2017) 012013 doi:10.1088/1742-6596/799/1/012013

Safety Analysis of Pb-208 Cooled 800 MWt Modified CANDLE Reactors

Zaki Su'ud[a] ,Nina Widiawati[a], H. Sekimoto[b], Artoto A[c].
[a]Nuclear and Biophysics Research Division, Faculty of Mathematics and Natural Sciences, Bandung Institute of Technology
[b]Tokyo Inst. of Technology (Emeritus Profesor), Japan
[c]Dept. of Physics, Jember National University, Jember, Indonesia

E-mail: szaki@fi.itb.ac.id

Abstract. Safety analysis of 800MWt Pb-208 cooled fast reactors with natural Uranium as fuel cycle input employing axial-radial combined Modiified CANDLE burnup scheme has been performed. The analysis of unprotected loss of flow(ULOF) and unprotected rod run-out transient overpower (UTOP) are discussed. Some simulations for 800 MWt Pb-208 cooled fast reactors has been performed and the results show that the reactor can anticipate complete pumping failure inherently by reducing power through reactivity feedback and remove the rest of heat through natural circulations. Compared to the Pb-nat cooled long life Fast Reactors, Pb-208 cooled reactors have smaller Doppler but higher coolant density reactivity coefficient. In the UTOP accident case the analysis has been performed against external reactivity up to 0.003dk/k. And for ULOHS case it is assumed that the secondary cooling system has broken. During all accident the cladding temperature is the most critical. Especially for the case of UTOP accident. In addition the steam generator design has also consider excess power which may reach 50% extra during severe UTOP case..

1. Introduction

As the respons to the TMI II and Chernobyl accident Innovative Nuclear Power Plants with inherent safety capability have been widely developed with Lead cooled fast reactors as one of them. The key factors for lead cooled fast reactors inherent safety capability are their negative reactifity feedbacks and their natural circulation capability. There are four important feedback in lead cooled fast reactors: Doppler, fuel axial expansion feedback, core radial expansion feedback, and coolant density feedback. In case of sodium cooled fast power reactors which more widely developed in many countries the coolant density reactivity feedback is positive except in some special cases. On the other hands, most lead cooled fast reactors have negative coolant density reactivity feedback. In large lead cooled fast reactor core it may be positive but the absolute value is not so large.

Pb-208 has low capture cross section in the fast region, low inelastic capture cross section in the fast energy region which give harder spectrum, potential to improve coolant void coefficient, and potential to improve neutronic performance of small long life liquid lead cooled fast reactors. In general the use of Pb-208 as coolant give better coolant density reactivity feedback than that of natural lead. It also give better criticallity than that of natural lead. Therefore the usage of Pb-208 as coolant in this study can be expected will contribute significantly to the overall safety performance.[1-2]

In this study, accident analysis has been performed for Pb-208 cooled 800 MWt Modified CANDLE based fast reactors[3-5]. The accident analysis include unprotected loss of flow accident (ULOF accident) and unprotected rod run-out transient overpower accident (UTOP accident). In this analysis coupled neutronic and thermal hydraulic analysis which include adiabatic model in nodal approach of time dependent multi-group diffusion equations has been adopted. The thermal hydraulic model include transient model in the core, steam generator, and related systems. Natural circulation based heat removal system is important to ensure inherent safety capability during unprotected accidents. Therefore the system similar to RVACS (reactor vessel auxiliary cooling system) is one of the important part to ensure the inherent safety feature of these reactors.[6-13]

2. Methodology[1,6-14]

The detail model and mathematical formulations for the current safety analysis can be obtained in the following references. Here the algorithm of the simulation will be discussed. After calculating effective macroscopic group constant then steady state multi-group diffusion calculation and steady state thermal hydraulic calculation are performed. The accident simulation begins with the accident initiator such as withdrawal of all control rods in case of UTOP accident and total loss of pumping power in the primary system for the case of ULOF accident. Then the calculation of total coolant flow-rate and flow distribution across the reactor core are performed which followed by the calculation of coolant and fuel temperature distribution, the calculation of energy and mass balance in the steam generator. Next, we solve the kinetic equation after calculating current feedback reactivities (Doppler, coolant density, core radial expansion and fuels axial expansion reactivity feedbacks) which then used in the calculation of amplitude and shape functions. The calculation is repeated by going back to the calculation of total coolant flow-rate and flow distribution across the core until reach the end of the simulation time.

3. Results and discussions

Table 1 shows main paramters for sample parameters for the current simulations. The reactor power is 800 MWt and this is a medium sized Modified CANDLE burn-up system.

Table 1 Main paramters used in this simulations

Parameter	Value
Power (MWth)	800
Core Geometry	2-D Cylinder
Refueling Periode(years)	10 years x10 batch
Fuel/cladding/coolant type	UN and PuN/SS316/Pb-208
Active core radius/height	117 cm/237.5cm
Reflector width (Pb-208)	70 cm
Fuel/structure/coolant volume fractions	60/12.5/27.5

The simulation results for Unprotected Loss of Flow (ULOF) accident case are shown in figures 1-4. Figure 1 shows coolant flow-rate in the core during unprotected loss of flow accident. Figure 2 shows power change with time during this accident. Fig. 3 shows The coolant decrease triggers imbalance between produced heat and coolant flow-rate causing temperature increase in coolant and fuel. Figure 3 shows coolant, cladding and fuel temperature change with time during the accident and finally Figure 4 shows reactivity feedback change with time during the accident. Temperature increase in coolant and fuel produces negative reactivity feedback which drive reactor power to decrease. After

400 seconds the coolant flow rate and the reactor power are approaching asymptotic levels and the total reactivity approaching zero. The maximum temperature in fuel and coolant are still far below safety limit.

The flow-rate decrease basically influenced by natural circulation level of the thermal hydraulic system of the NPP. About 200s after the accident begins, the core and primary side SG flow-rate ave approached natural circulation levels and after 400 seconds the flow-rate change become relatively slow. In case of the reactor core power level, it also decreases following the decrease of the core flow-rate. However the asymptotic level is still at higher percentage compared to that of flow-rate.

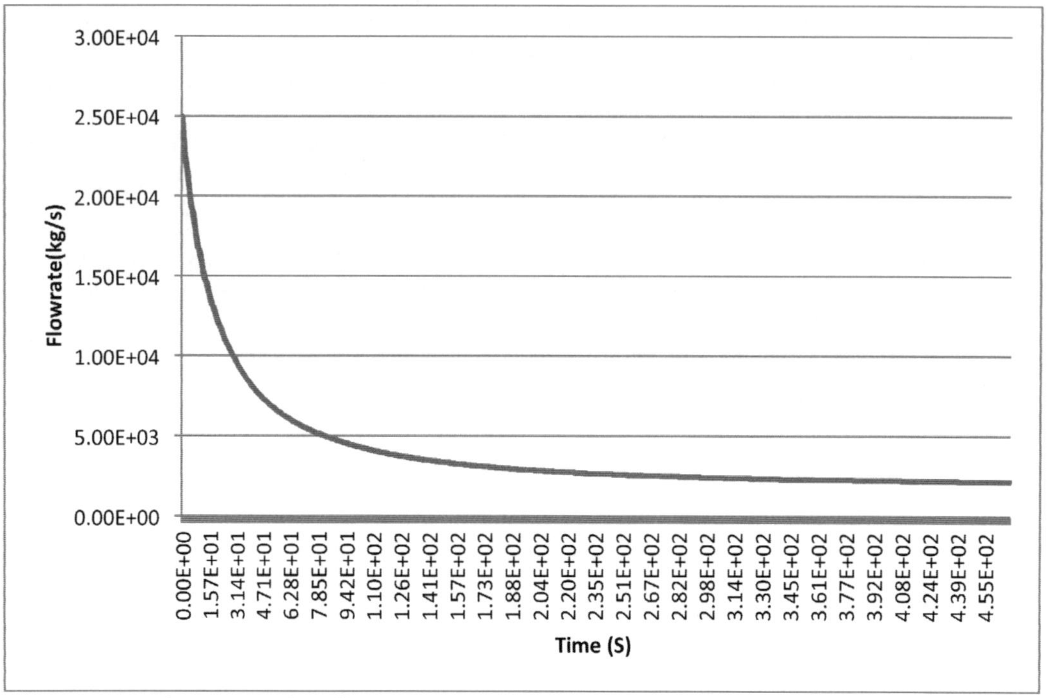

Figure 1 Coolant flow-rate change with time during loss of flow accident

After the decrease of the coolant flowrate, the fuel, cladding and cooolant temperatures initially increase and after reaching the peak then decrease and move toward new equilibrium level. The fuel temperature experience more significant reduction compared to the coolant and cladding temperature. This situation can be explained due to relatively large coolant-center fuel temperature different in the fuel pin. Due to the change of the fuel to coolant temperature difference during the accident the position of maximum fuel, cladding and temperatures are basically changing with the time and may not in the same positions for each coolant maximum temperature, cladding maximum temeprature, and fuel maximum temperature.

The reactivity feedbacks move in more complex situations. The Doppler and fuel axial expansion at the beginning give strong negative feedback, however, after reaching maximum temperature then their absolute values decerase and finally move toward positive direction. On the other hand, the core radial expansion and the coolant density reactivity feedback are definitely negative. However their values also decrease after reaching their maximum peak(absolute).

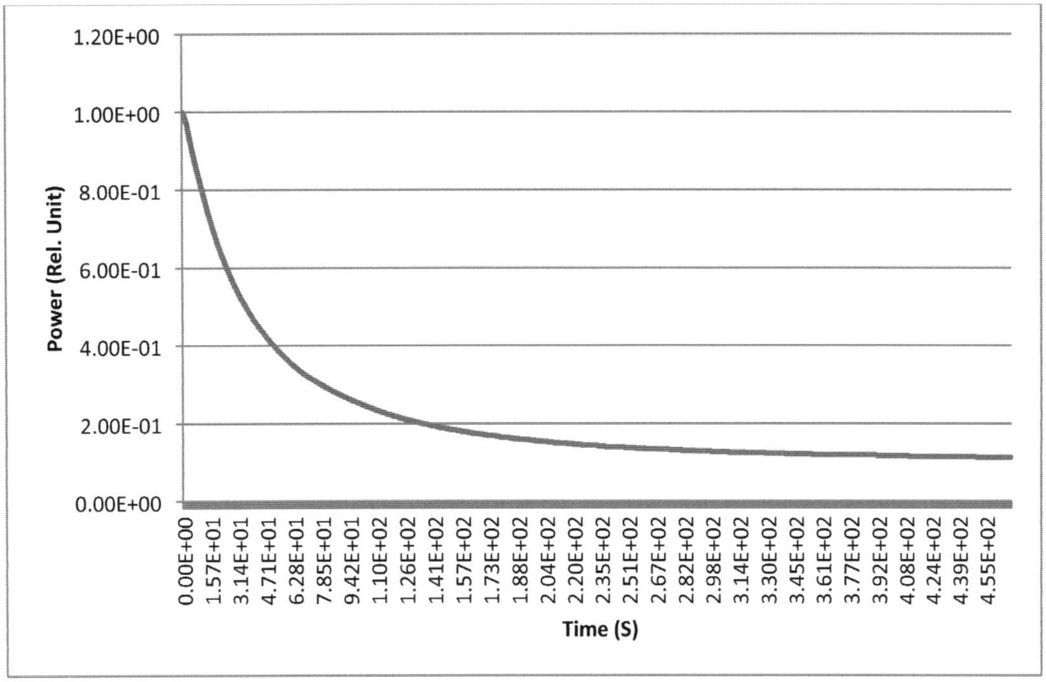

Figure 2 Power change with time during loss of flow accident

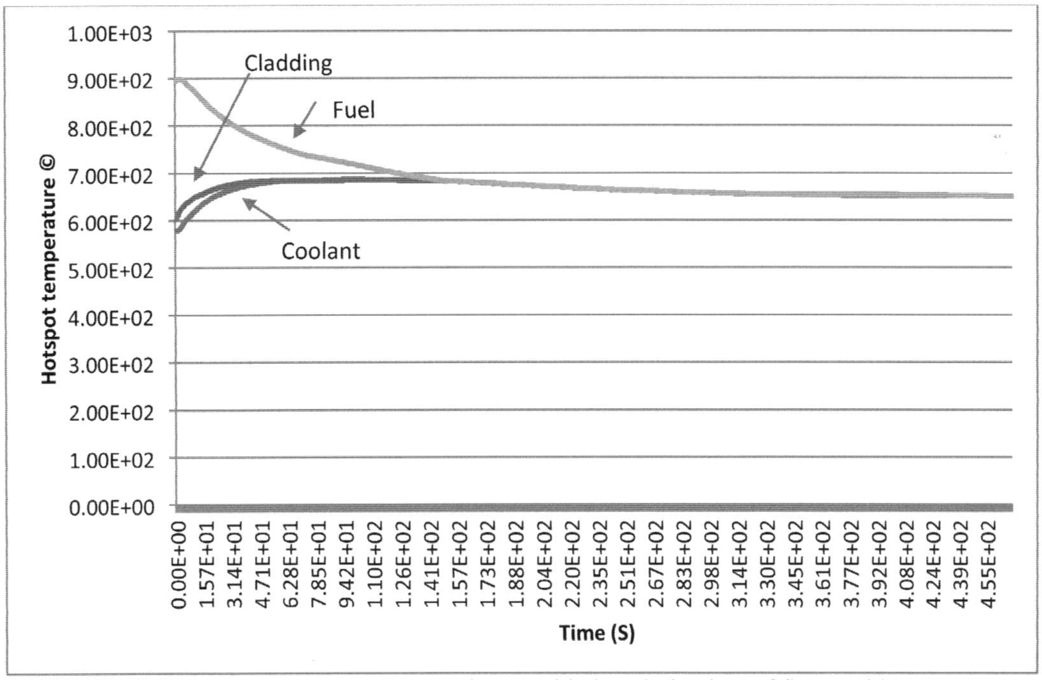

Figure 3 Hot spot temperatures change with time during loss of flow accident

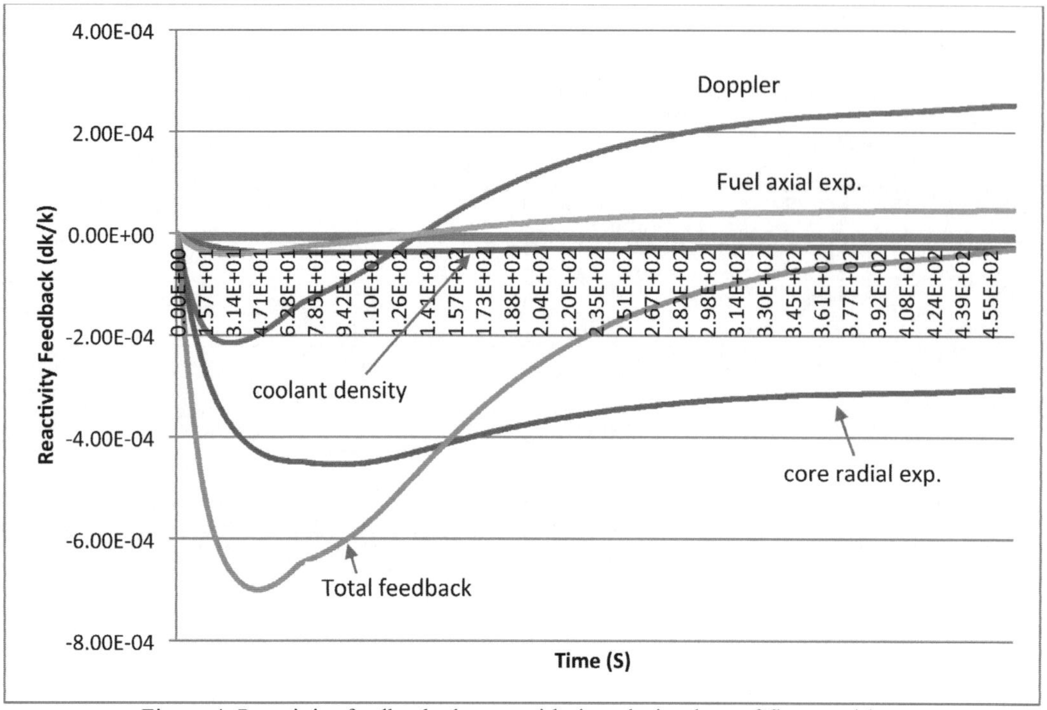

Figure 4 Reactivity feedback change with time during loss of flow accident

The accident analysis simulation results for Unprotected rod run-out Transient Over Power (UTOP) accident are shown in figures 5-8. As the accident intitator, the positive external reactivity triggers power increase which then causes temperature increase in coolant, cladding and fuel. Temperature increase in coolant, cladding and fuel produces negative reactivity feedback which is used to gradually compensate external positive reactivity. After 200 seconds the power level approaching asymptotic levels and the total reactivity approaching zero. Doppler and Core radial expansion feedback play dominant role in UTOP accident. The temperature increase in fuel and coolant is still far below safety limit.

There are overshoot pattern of power following fast withdrawal of the control rod. Therefore this pattern influence the pattern of hotspot fuel, cladding and coolant temperature change with time and also the feedback. The most important difference between ULOF and UTOP cases are the fact that in the UTOP accident the fuel, cladding and coolant temperatures are continuously increase till approaching the asymptotic new power levels while in the ULOF case there are peak values of fuel coolant and cladding temperature as shown in Figures. 3 and 7. In the UTOP case the power is also increases till reaching asymptotic level while in the ULOF case decreases and then approaching asymptotic new power level as shown in Figs.3 and 6. In case of the UTOP accident the position change of maximum temperature for coolant, cladding and coolant are not significant.

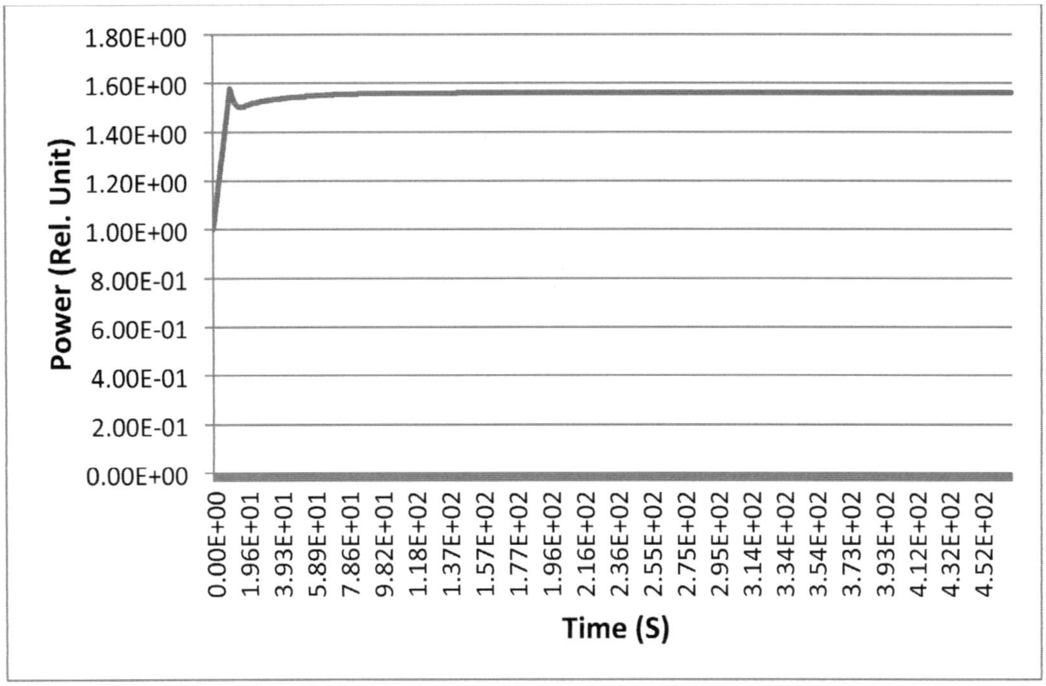

Figure 5 Power change with time during UTOP accident

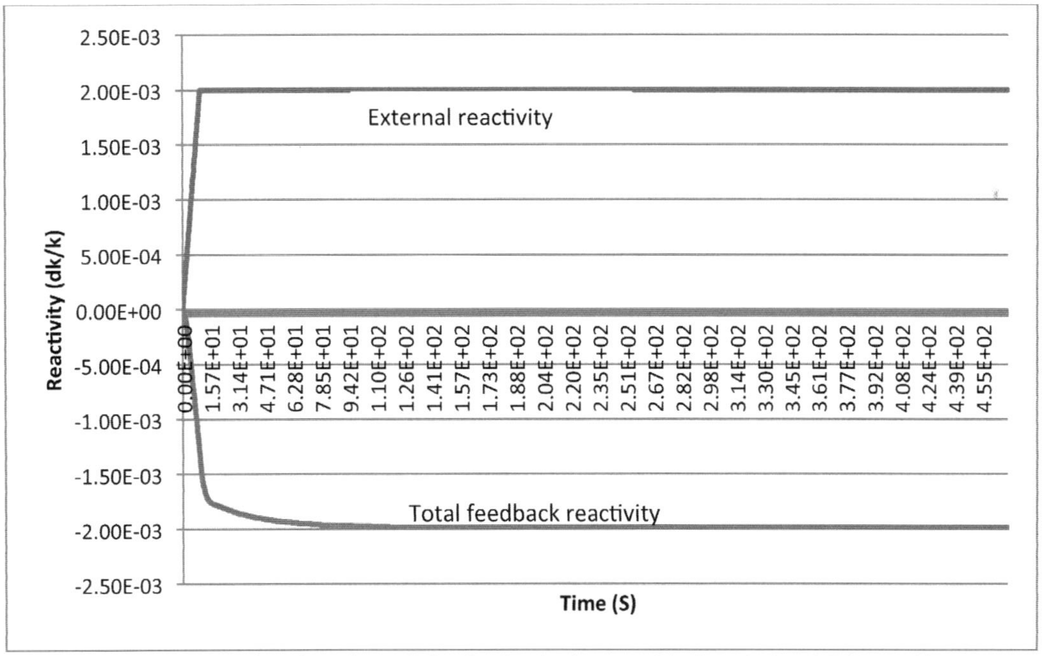

Figure 6 External and total reactivity feedback change with time during UTOP accident

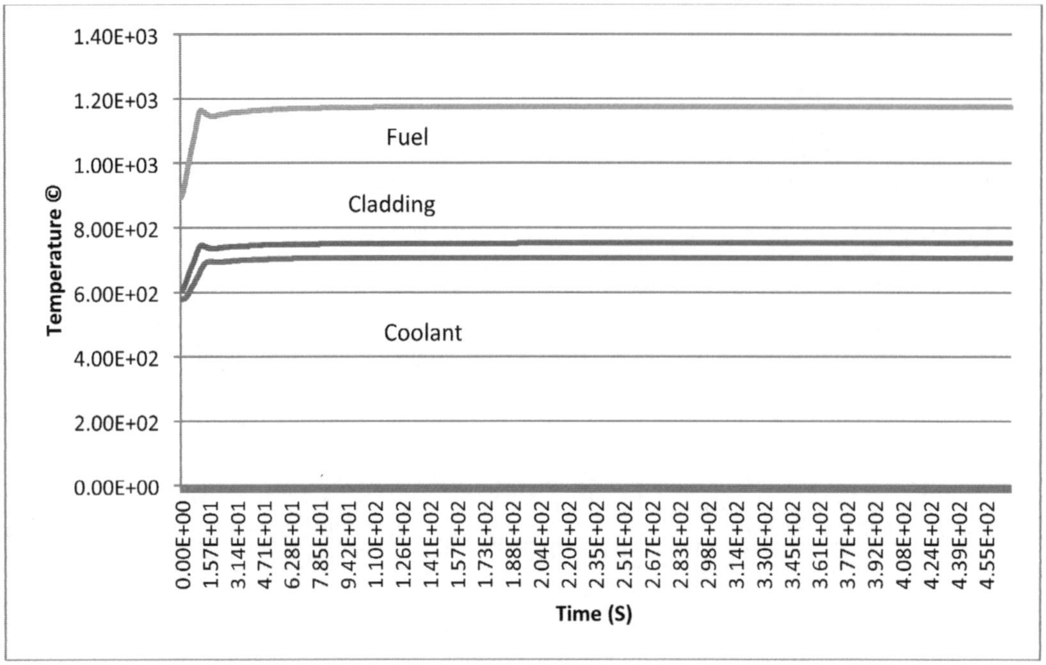

Figure 7 Hot spot temperatures chhange with time during UTOP accident

Figure 8 shows teh change of external and feedback reactivities during UTOP accident simulations. As shown in this figure, the new equilibrium conditions can be reached after total feedback reactivity can completely compensate external reactivity. From figure 8 it is shown that after 400 seconds from the accident begin the asymptotic level have been approached.

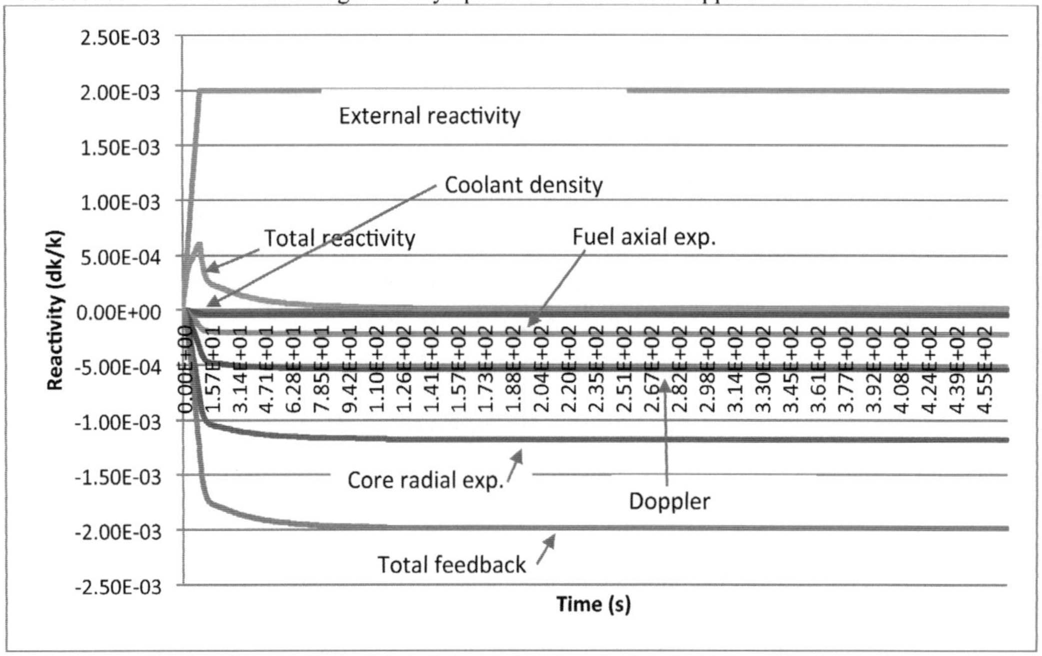

Figure 8 Reactivity change with time during UTOP accident

IOP Conf. Series: Journal of Physics: Conf. Series **799** (2017) 012013 doi:10.1088/1742-6596/799/1/012013

4. Conclusion

Safety analysis simulations have been performed for Pb-208 cooled 800MWt Modified CANDLE fast reactors especially against unprotected loss of flow (ULOF) and unprotected rod-run out transient overpower (UTOP). Pb-208 Cooled Small Modified CANDLE Reactors can survive ULOF and UTOP inherently. Natural circulation plays important role in the ULOF accident. Core radial expansion and Doppler reactivity feedback plays important role in the ULOF and UTOP accident.

References

[1] Zaki S,, 2008, Progress in Nuclear Energy, Neutronic Performance Comparation of MOX, Nitride and Metallic fuel based 25-100 MWe Pb-Bi Cooled Long Life Fast Reactors without on site Refuelling, Vol. 50, p. 276-278
[2] Zaki S, 2008, Progress in Nuclear Energy, Safety Performance Comparation of MOX, Nitride and Metallic fuel based 25-100 MWe Pb-Bi Cooled Long Life Fast Reactors without on site Refuelling, Vol. 50, p. 157-162
[3] Zaki Su'ud and H Sekimoto, 2010, IJNEST, Design study of long-life Pb-Bi cooled fast reactor with natural uranium as fuel cycle input using modified CANDLE burn-up scheme, Vol 5, No. 4 , p.347-368
[4] Su'ud, Z., and Sekimoto, H., 2012, IJNEST, Design study of medium-sized Pb-Bi cooled fast reactors with natural uranium as fuel cycle input using modified CANDLE burn-up scheme, Vol. 7 (1) , pp. 23-44.
[5] Zaki Su'ud and H Sekimoto, 2013, Annals of Nuclear Energy, The prospect of gas cooled fast reactors for long life reactors with natural uranium as fuel cycle input, Vol. 54 , pp. 58-66.
[6] S.K. Cheng and N.E. Todreas, 1986, Hydrodynamic models and correlationx for bare and wire-wrapped hexagonal rod bundles, Nucl. Eng. Des. 92, p 227.
[7] Zaki Su'ud and H. Sekimoto, 2010, IJNEST, Local Blockage analysis of Lead-Cooled Next Generation Nuclear Power Reactors, Vol. 5, No.2
[8] J.G. Guppy, 1993, Super System Code, NUREG/CR-3169.
[9] S. Kakac, R.K. Shah and W.Aung, 1987, Handbook of Single Phase Convective Heat Transfer, Wiley, pp.7.1-7.62.
[10] K.O. Ott and R.J. Neuhold, 1985, ANS, Nuclear Reactor Dynamics.
[11] H.Sekimoto and S. Zaki, 1995, Nucl. Technol. , Design study of lead and lead-bismuth cooled small long-life nuclear power reactors using metallic and nitride fuel, 109,p. 307-313.
[12] Zaki S. dan Sekimoto,H., 1996, Nucl. Eng. And Design, Accident Analysis of Lead or Lead-Bistmuth Cooled Small Safe Long-Life Fast Reactor using Metallic or Nitride Fuel, Vol. 162, Elsevier Science, pp.205-222.
[13] Zaki S., 1998, Progress of Nuclear Energy, Comparative Study on Safety Performance of Nitride Fueled Lead-Bismuth Cooled Fast Reactor with Various Power Level", Vol.32, No. ¾,, 571~577
[14] N. Nakagawa and K.Tsuchihashi, 1984, JAERI Memo, SLAROM: a code for cell homogenization calculation of fast reactor, (Japan Atomic Energy Research Institute, Ibaraki), JAERI M-1294

Synthesis of Al_2O_3 Nanoparticles from Local Bauxite for Water- Al_2O_3 Nanofluids egy

Dani Gustaman Syarif*, Djoko Hadi Prajitno, Efrizon Umar

Center for Applied Nuclear Science and Technology (PSTNT)
National Nuclear Energy Agency (BATAN)
Jl. Tamansari 71, Bandung 40132, Indonesia

*Email: danigus@batan.go.id

Abstract. A study on synthesis and characterization of Al_2O_3 nanoparticles from bauxite for water-Al_2O_3 nanofluids as an alternative coolant for nuclear and non-nuclear applications has been done. The Al_2O_3 nanoparticles were synthesized by heating AlOOH as a precursor derived from local bauxite. The Al_2O_3 nanoparticles were dispersed in water and ultrasonicated to produce nanofluids. XRD data showed that the Al_2O_3 nanoparticles crystallized in gamma alumina with crystallite size of 4.5 nm (Debye Scherrer method). Surface area of the Al_2O_3 nanoparticles was 195 m^2/gram. Data of TEM showed that the particle size was smaller than 10 nm. According to zeta potential data, the nanofluids were stable at neutral pH of 7.3 with zeta potential of 28-51 mV. The height of the nanofluid surface decreased about 13 % after 6 days. In addition, the CHF of the water-Al_2O_3 nanofluids produced in this study increased about 55-161 % compared to that of water.

1. Introduction

There is a need to improve heat transfer capability of conventional fluid such as water, ethylene glycol, and oil as reported by some researchers [1-6]. The improvement of the heat transfer capability of the conventional fluids is to increase the efficiency of some systems based on heat transfer such as automotive and nuclear reactor, and of some equipments and processes such as electronics, and metal forming. Introducing nanofluids, as a product of nanotechnology, to heat transfer system to replace the conventional fluids is intensively studied over decades [1-5]. The nanofluids are stable suspensions consisting dispersed nanoparticles with particle size of 1-100 nm and base fluids such as water, ethylene glycol, and oil [2,6].

Some characteristics are required by the nanofluids before application and one important characteristic of the nanofluids is critical heat flux (CHF) i.e. the maximum heat flux attainable from the nucleate boiling. An enhanced CHF is advantageous for increasing the safety margin of the thermal system, and to design compact and efficient cooling systems. The nanofluid with good CHF is required for some applications such as ECCS (Emergency Core Cooling System) and RVCS

(Reactor Vessel Cooling System) mainly required in nuclear reactor system, quenching in metal heat treatment, metal cutting and machining.

Some works dealing with CHF of the nanofluids have been reported [1,3, 7-11]. It was reported that some nanofluids had good potential as fluid having large CHF namely water-CuO [3], water-ZrO_2 [3], water-Fe_3O_4 [11,12], water-ZnO [13], and water-Al_2O_3 [1, 3, 10,11]. Nanofluid of water-Al_2O_3 is interesting in heat transfer application due to special characteristic of Al_2O_3 namely low density, low neutron absorption coefficient, and chemical inertness that is good to avoid corrosion problem. Although the nanofluid of water-Al_2O_3 has been studied in relation with the CHF, however, the CHF of the nanofluids with neutral pH in the previous studies was not reported and discussed. It was known that some nanofluids had acidic and basic pH. The neutral pH is important to avoid corrosion problem.

In order to increase the added value of mineral abundant in Indonesia especially bauxite, in this work, the water-Al_2O_3 nanofluid was prepared by utilizing Al_2O_3 nanoparticles derived from local bauxite. The nanofluids having neutral pH were prepared and characterized. The characteristics of the nanofluids were discussed in relation with the CHF and heat transfer.

2. Experimental Method

Synthesis and Characterization of Al_2O_3 Nanoparticle

An amount of bauxite and NaOH was mixed and ground. The mixture was suspended in water, and then the suspension was heated in an autoclave at about 175°C for 3 hours. After heating, the suspension was cooled down. Precipitate of red mud was separated from filtrate by filtration. The filtrate was precipitated by adding HCl. The precipitate was separated by filtration, and washed using water. The precipitate was heated at 200°C to form AlOOH.

An amount of AlOOH was heated at 900°C for 3 hours to get Al_2O_3 nanoparticles. The Al_2O_3 nanoparticles were analyzed using XRF to know the chemical composition, and then they were subjected to XRD analysis to know their crystal structure and crystallite size (Debye Scherrer method). Nanoparticle size was measured using a TEM belongs to TEM laboratory of Chemistry Department of UGM, and surface area of the nanoparticles was measured using a surface area meter of quantachrome NOVA 2200.

Synthesis and Characterization of Water-Al_2O_3 Nanofluids

The Al_2O_3 nanoparticles of 0.5, 1, 1.5, and 2 grams were suspended in 100 ml water, and ultrasonicated using an ultrasonic bath for 2 hours to produce nanofluids. After calculation, the concentrations of nanofluids are 0.125, 0.250, 0.375, and 0.5 vol %. Visually, a typical nanofluid was observed time to time. The height of the nanofluid surface was also measured time to time. Zeta potential of the nanofluids was measured using Zetasizer from Malvren. Meanwhile, viscosity of the nanofluids was measured using a vibro viscometer SV-10. CHF of the nanofluids was measured using a method found in works of Lee, et al.[11] and Hiswankar and Kshirsagar [13] by utilizing Cu wires with diameter of 0.2 mm.

3. Results and Discussion

Visual appearance of the Al_2O_3 nanoparticles is depicted in Fig. 1 and its XRD pattern can be seen in Fig. 2. Chemical composition of Al_2O_3 nanoparticles analyzed using XRF is depicted in Table 1. From Fig. 2, it can be seen that the Al_2O_3 nanoparticle crystallizes in gamma alumina. Peaks belong to the pattern of the Fig. 2 are broadening indicating very small crystallite of the powder. Based on a calculation using Debye Scherrer method [14], the average crystallite size of the nanoparticles was 4.5 nm. Specific surface area associated with the nanoparticles is 195 m^2/gram. This surface area is large enough. When particle size of the nanoparticles is estimated from the specific surface area by using

IOP Conf. Series: Journal of Physics: Conf. Series **799** (2017) 012014 doi:10.1088/1742-6596/799/1/012014

equation (1) [15], it is known that particle size is 7.75 nm. This value is in the same range with that calculated from the XRD pattern using Debye Scherrer method. The little difference is caused by agglomeration of the nanoparticles.

$$d = \frac{6000}{\rho . As} \quad \text{..............................(1)}$$

Where, d = particle size (nm), ρ = theoretical density of nanoparticles (g/cm^3), and As = specific surface area (m^2/g).

Figure 1. Appearance of the Al_2O_3 nanoparticles.

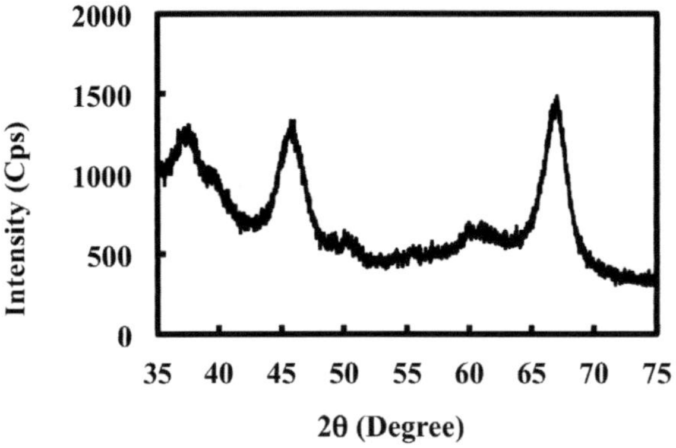

Figure 2. XRD pattern of the Al_2O_3 nanoparticles.

Table 1. Chemical composition of the Al_2O_3 nanoparticles.

No.	Component	Concentration (Weight %)
1	Al_2O_3	99.443
2	SiO_2	0.230
3	TiO_2	0.002
4	Fe_2O_3	0.035
5	CaO	0.061
6	MgO	0.134
7	Na_2O	0.054
8	K_2O	0.026
9	SO_3	0.016

Fig. 3 is a TEM image of the Al_2O_3 nanoparticles. According to Fig. 3, the Al_2O_3 nanoparticles are small with size about 7 nm and tend to form agglomerates. Al_2O_3 powder forming agglomerate with size about 80 nm contains individual small particle was also synthesized by Karim et al. [16] using sol gel utilizing urea as the chelating agent. Surface area of the powder of Karim et al. [16] is only 4.4 m^2/g. This value is much smaller than that of the powder in this work of 195 m^2/g. The difference may be due to different treatment.

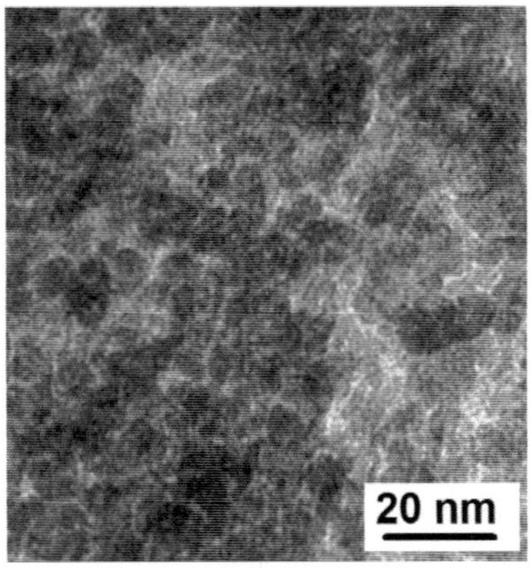

Fig. 3. TEM image of the Al_2O_3 nanoparticles.

A nanofluid with Al_2O_3 concentration of 0.25 vol % that was prepared using the Al_2O_3 nanoparticle synthesized in this study is shown in Fig. 4. The observation was done only to the nanofluid with pH 7.3 because the most applicable nanofluid is that having neutral pH (with no corrosion problem). The height of the surface of the nanofluid decreases time to time. Quantitatively the decrease of the height of the nanofluid surface is shown in Fig. 5. The height of the nanofluid surface decreases about 13 % of the initial height of the nanofluid surface after 6 days observation. Compared to the water-Al_2O_3 nanofluid in the previous study [17], the stability of this nanofluid is better. In the previous study [17], the height of the nanofluid decreased 20% after 6 days observation. The stability of the water-Al_2O_3 nanofluid in this study is also better than that of the same nanofluid found in a literature [18]. The good stability may be due to small particle size and large surface area.

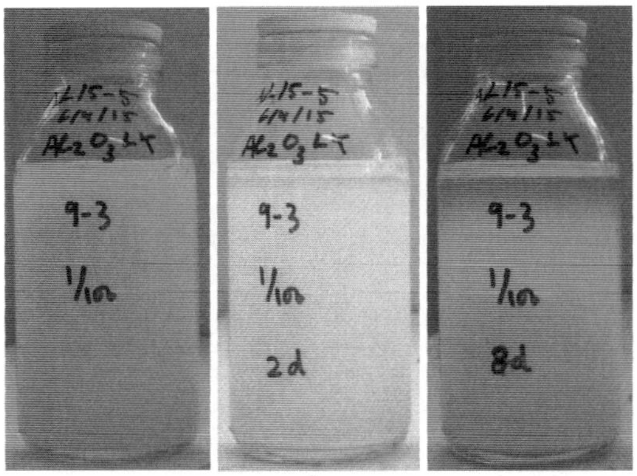

a b c

Fig.4. Water-Al_2O_3 nanofluids at different observation times (a. Initial, b. 2 days, c. 8 days).

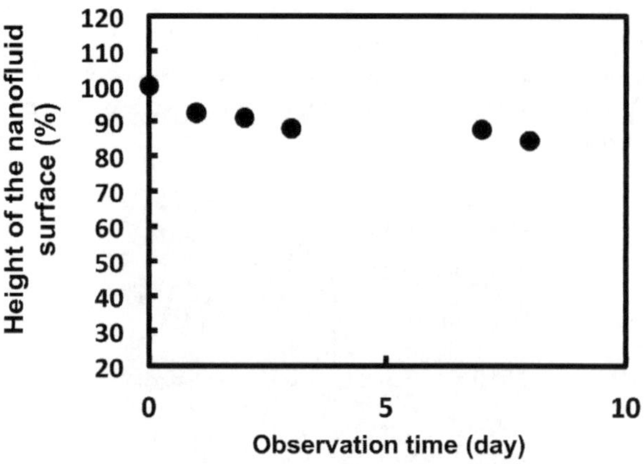

Fig. 5. The relation between the height of the nanofluid surface and observation time. Concentration of Al_2O_3 is 0.25 vol% and pH 7.3.

Fig. 6 shows zeta potential of the water-Al_2O_3 nanofluids. It can be seen that the zeta potential changes is relatively independent on the change of concentration of the nanoparticles. The nanofluid is relatively stable when zeta potential is larger than 25 mV or smaller than -25 mV. According to this ctriterium, from Fig. 6, it can be seen that the nanofluids with different concentrations prepared in this study are stable at pH 7.3.

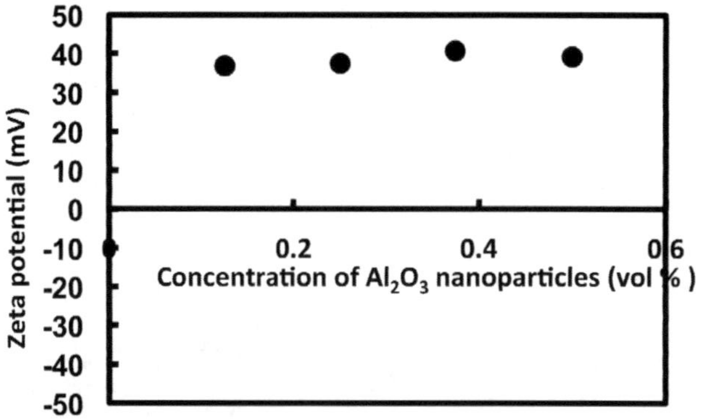

Fig.6. Zeta potential of the water-Al_2O_3 nanofluids.

Viscosity data for the nanofluid with pH 7.3 depicted in Fig. 7. The viscosity of the water-Al_2O_3 nanofluid with pH of 7.3 increases linearly with concentration of Al_2O_3 nanoparticles within the range of concentration. Compared to Einstein prediction, the gradient of increase is larger. It means that the viscosity increase of the nanofluids as the increase of concentration does not obey the Einstein prediction. This difference happens because the nanofluids are more complicated than the suspension used for the Einstein prediction. The nanofluid does not simply refer to a liquid-solid mixture. The characteristics such as pH, zeta potential, and Brownian motion effect that were not considered during formulation of the Einstein prediction are very important for the nanofluids.

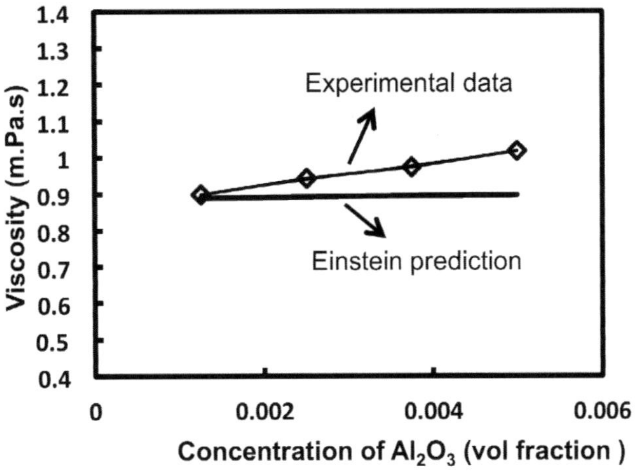

Fig.7. Viscosity of the water-Al_2O_3 nanofluid with pH 7.3.

Fig. 8 shows the increase of CHF as function of nanoparticles concentration. The increase of CHF increases almost linearly from 55 to 161 % with nanoparticles concentration. This means that heat transfer from Cu wire to the nanofluids goes well when the nanoparticles concentration increases. The mechanism of the increase of CHF in the nanofluid is due to nanoparticles coating on Cu wire surface [8]. The heated surfaces are changed by the deposition of nanoparticles. The deposited nanoparticles generated a porous layer on the Cu wire, which then improve the wettability of the wire. The heat transfer becomes better when the nanoparticles concentration increases. The degree of CHF enhancement increases as the amount of deposited nanoparticles increases. So why the CHF is more enhanced at higher concentrations of the nanoparticles. By considering the characteristics of the nanofluids prepared in this study, the nanofluids have possibility to be applied in nuclear reactor as fluid or coolant for ECCS (Emergency Core Cooling System) and RVCS (Reactor Vessel Cooling System), and in automotive, electronics, and machining and cutting process as also the coolant.

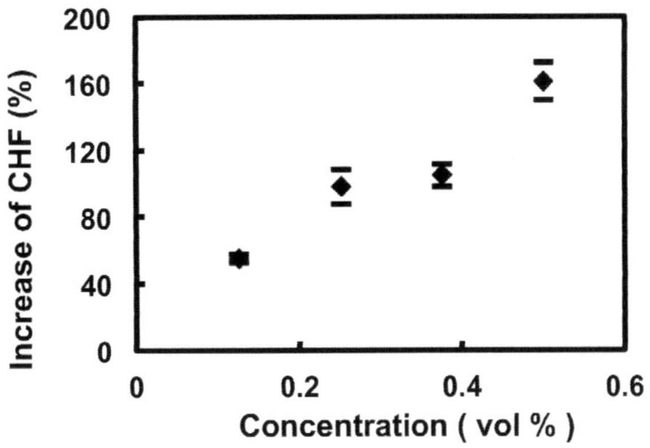

Fig. 8. Increase of CHF of the water-Al_2O_3 nanofluids as function of Al_2O_3 concentration.

4. Conclusion

Al_2O_3 nanoparticle with crystallite size of 4.5 nm has been well synthesized from AlOOH as precursor derived from local bauxite by calcination. Although the nanoparticle forms agglomerate, the nanofluid made of this nanoparticle has relatively good stability at neutral pH (about 7), the pH that is required to avoid corrosion problem. The CHF of the nanofluids is larger than that of water, and increases from 55 to 161 % with Al_2O_3 concentration from 0.125 to 0.5 vol %. The nanofluids have possibility to be applied in nuclear application as fluid or coolant for ECCS (Emergency Core Cooling System) and RVCS (Reactor Vessel Cooling System), and in non-nuclear applications as coolant for automotive, electronics, metal heat treatment (for quenching), and machining and cutting process.

Acknowledgments

Authors would like to acknowledge the financial support from BATAN through DIPA program year 2014 and 2015. Special thanks to Mr. M. Yamin for his help during synthesis of Al_2O_3.

References

[1] R. A. Taylor, P.E. Phelan, Pool boiling of nanofluids: Comprehensive review of existing data and limited new data, *Int. J. Heat Mass Transfer* 52, 53339, 2009.

[2] R. Saidur, K.Y. Leong, H.A. Mohammad, A review on applications and challenges of nanofluids, *Renew. Sustain. Energy Rev.* 15, 1646, 2011.

[3] H. Kim, Enhancement of critical heat flux in nucleate boiling of nanofluids: a state-of-art review, *Nanoscale Res. Lett.* 6, 415, 2011.

[4] W. Yu et al., Comparative review of turbulent heat transfer of nanofluids, *Int. J. Heat Mass Transfer* 5, 5380, 2012.

[5] I. Nkurikiyimfura, Y. Wang, Z. Pan, Heat transfer enhancement by magnetic nanofluids-A review, *Renew. Sustain. Energy Rev.* 21, 548, 2013.

[6] O. Mahian et al., A review of the applications of nanofluids in solar energy, *Int. J. Heat Mass Transfer* 57, 582, 2013.

[7] S. W. Lee et al., Critical heat flux enhancement in flow boiling of Al_2O_3 and SiC nanofluids under low pressure and low flow conditions, *Nucl. Eng. Tech.* 44(4), 429, 2012.

[8] Ho Seon Ahn, M. H. Kim, A Review on critical heat flux enhancement with nanofluids and surface modification, *J. Heat Transfer* 134,1, 2012.

[9] T. Okawa, M. Takamura, T. Kamiya, Boiling time effect on CHF enhancement in pool boiling of nanofluids, *Int. J. Heat Mass Transfer* 55, 2719, 2012.

[10] J. Y. Jung et al., The study on the critical heat flux and pool boiling heat transfer coefficient of binary nanofluids (H_2O/LiBr D Al_2O_3), *Inter. J. Refrigeration* 36, 1051, 2013.

[11] J. H. Lee, T. Lee, Y.H. Jeong, The effect of pressure on the critical heat flux in water-based nanofluids containing Al_2O_3 and Fe_3O_4 nanoparticles, *Int. J. Heat Mass Transfer* 61, 432, 2013.

[12] D. G. Syarif, D.H. Prajitno, Synthesis and characterization of Fe_3O_4 nanoparticles and water-Fe_3O_4 nanofluids, *Proceedings of the International Forum on Strategic Technology (June 2-3, 2015, Kuta,Bali, Indonesia)*,342, 2015.

[13] S. C. Hiswankar, J. M. Kshirsagar, Determination of critical heat flux in pool boiling using ZnO nanofluids, *Int. J. Eng. Res. & Tech. (IJERT)* Vol. 2 (7), 2091, 2013.

[14] K. S. Rathore et al., Structural and optical characterization of chemically synthesized ZnS nanoparticles, *Chalcogenide Lett.* 5 (6), 105, 2008.

[15] B. Akbari, M. P. Tavandashti, M. Zandrahimi, Particle size characterization of nanoparticles-A practical approach, *Iranian J. Mater. Sci. Eng.* Vo. 8 (2), 48, 2011.

[16] M. R. Karim et al., Synthesis of gamma-alumina particles and surface characterization, *Open Coll. Sci. J.* 4, 32, 2011.

[17] D.G. Syarif, D.H. Prajitno, Synthesis and characterization of Al_2O_3 nanoparticles and water-Al_2O_3 nanofluids for nuclear reactor Coolant, *Adv. Mater. Res.* 1123, 270, 2015.
[18] R. A. Bhogare, B. S. Kothawale, A review on applications and challenges of nanofluids as coolant in automobile radiator, *Inter. J. Sci. Res. Pub.* 3(8), 1, 2013.

IOP Conf. Series: Journal of Physics: Conf. Series 799 (2017) 012015 doi:10.1088/1742-6596/799/1/012015

Radiological dose in Muria peninsula from SB-LOCA event

Sunarko[1*], **Zaki Suud**[2]

[1]BATAN, Jl. Kuningan Barat, Jakarta 10270, Indonesia
[2] ITB, Jl. Ganesha 10, Bandung 50254, Indonesia

Email: sunarko@batan.go.id

Abstract. Dose assessment for accident condition is performed for Muria Peninsula region using source-term from Three-Mile Island unit 2 SB-LOCA accident. Xe-133, Kr-88, I-131 and Cs-137 isotopes are considered in the calculation. The effluent is assumed to be released from a 50 m stack. Lagrangian particle dispersion method (LPDM) employing non-Gaussian dispersion coefficient in 3-dimensional mass-consistent wind-field is employed to obtain periodic surface-level concentration which is then time-integrated to obtain spatial distribution of ground-level dose. In 1-hour simulation, segmented plumes with 60 seconds duration with a total of 18.000 particles involved. Simulations using 6-hour worst-case meteorological data from Muria peninsula results in a peak external dose of around 1.668 mSv for low scenario and 6.892 mSv for high scenario in dry condition. In wet condition with 5 mm/hour and 10 mm/hour rain for the whole duration of the simulation provides only minor effect to dose. The peak external dose is below the regulatory limit of 50 mSv for effective skin dose from external gamma exposure.

1. Introduction

The capability to estimate radioactive dose dispersion in a course of an accident in a nuclear facility is important to support emergency planning activities and ensure public safety from ionizing radiation exposure. In nuclear power plant site study stage, the information contained in this analysis can be used as an input in emergency response planning, radiation protection activities, and nuclear power plant siting and design.

A computer code named MODAL has been developed based on Lagrangian Particle Dispersion Method[2][12]. The code is capable of handling spatial and temporal variation in topography, wind and source-term. The inputs for this code include topography, surface wind station data and source-term which are usually available during the site study phase through survey, monitoring or other means of data collection.

The Three-Mile Island unit 2 (TMI-2) nuclear accident occurred in March 28, 1979. It was a scale-5 in International Nuclear Event Scale which resulted in some damage to the reactor core and release of radioactive substance to the environment. According to the US-Nuclear Regulatory Committee, this is the most serious accident in the U.S. commercial nuclear power plant operating history. A number of literatures have discussed and reconstructed the amount of radioactive released to the atmosphere for several isotopes such as I-131, Cs-137, Kr-88 and Xe-133. Haste et.al.,[5] through simulation with MELCOR/MACCS proposed a total mass of about 0.1 kg of noble gasses and 2.5×10^{-7} kg of CsI. The radionuclide is assumed to be released in a 6-hour period from seconds 8.500 to 30.000 after the accident started. Meteorological data in the model[5] was taken from Surry PWR which is about 200 miles away from TMI-2 site. The result of probabilistic analysis showed that average total calculated individual dose is less than 0.2 mSv at 95% probability and 0.4 mSv at 99% probability. Peak average individual dose are 0.5 and 1.1 mSv for 60 m and surface release height respectively. Jaffe (USNRC) is quoted to have estimated a peak of 0.7 mSv with a release of 2.5 million Ci of noble gasses and 15 Ci of I-131. Gudiksen and Dickerson [6] found about 2.4-10 million Curies of noble gasses with a majority of Xe-133 and 14 Curie of I-131 and 2.6 Curie of I-133. McColl and Prosser [7] found about 62 thousand Curie of Kr-88, 13.5 Ci of I-131, and 8.38 million Curie of Xe-133. The source term used in this paper is summarized in **Table 1**.

Table 1. Source-term

	Source strength rate (Bq/sec)				
Radionuclide	Haste et al.	McColl et al.	Gudiksen and Dickerson	NRC	Max.
Cs-137	2.27E+04	-	-	-	2.27E+04
I-131	3.12E+07	2.31E+07	2.40E+07	-	3.12E+07
Kr-88	-	1.06E+08	-	-	1.06E+08
Xe-133 (low)	-	1.44E+10	4.11E+12	4.28E+12	4.28E+12
Xe-133 (high)	-	-	1.71E+13	-	1.71E+13

This paper aims to assess radioactive dose imposed to the Ujung Lemahabang area in Muria Peninsula by a small-break loss-of-coolant accident (SB-LOCA) such as the Three-Mile Island unit 2 in the USA. Deterministic analysis is performed to obtain spatial distribution of dose in the case of sustained wind condition or in the worst-case meteorological condition.

2. Theory/calculation

Lagrangian Particle Dispersion Method (LPDM) is run by tracking a number of particles in a flow field. LPDM uses stochastic differential equation to explain the similar process as the advection-diffusion equation in Lagrangian framework[9].

Atmospheric turbulence is inherently stochastic and many researchers proposed an approach based on the statistical nature of turbulence. In stochastic representation, marked particles undergo advection process by wind at a certain speed and at the same time experiencing random movement simulating turbulent fluctuation. Average distribution of particles is determined by averaging particle paths. Since every particle moves independently, simultaneous handling of particles is unnecessary and therefore requires small computer memory [1]. The accuracy of this method increases with smaller computational volume.

The advection process of particles is provided by a diagnostic 3-dimensional wind field based on the Mass consistent method (MCM) proposed by Sasaki in 1958 and implemented by Sherman[10]. MCM employs minimization of functional in equation (1).

$$E(u,v,w,\lambda) = \int_V \left[\alpha_1^2(u-u^0)^2 + \alpha_1^2(v-v^0)^2 + \alpha_2^2(w-w^0)^2 + \lambda\left(\frac{\partial u}{\partial x} + \frac{\partial v}{\partial y} + \frac{\partial w}{\partial z}\right)\right] dxdydz \tag{1}$$

In this model, it is assumed that release rate, wind speed and wind direction are constant during the hour. Topography and meteorological data are used to construct a 3-dimensional wind field using mass-consistent method[10]. The wind field takes part in the advection process of the fictitious particles. The dispersion part is run using spatially variable empirical dispersion coefficient constructed based on the atmospheric stability condition and depth of boundary layer. These data are assumed because literature data is not available. Radioactive decay and wet deposition in case of rain is accommodated in the model.

Wet deposition of radioactive material, or tropospheric particles in general, is primarily accomplished by two mechanisms, namely 'rainout' and 'washout'. Rainout is the process of particles becoming entrained into cloud droplets, either by nucleation or by scavenging, and are subsequently removed from the atmosphere along with cloud water during precipitation. Washout is particle scavenging by falling precipitation. The scavenging process of washout consists of repeated exposures of particles and gases to cloud or precipitation elements with some chance of accretion by the elements for each exposure[8]. In the simulation, wet deposition is caused by washout because the radioactive plume is below cloud.

Wet deposition can be assumed as the decrease of radioactivity level in the atmosphere from the presence of rain. The decrease can be estimated using exponential form in equation (2) and coefficients provided in **Table 2** [11]:

$$\frac{dq_n}{dt} = -\Lambda q_n \quad (\Lambda = \alpha\gamma^{\beta}) \tag{2}$$

q_n is particle radioactivity, Λ (sec-1) is washout coefficient, α and β are factors for washout coefficient for a certain radionuclide and γ is the rainfall rate (mm/hour).

Table 2. Factors for washout coefficient Λ

Isotope	α	β
I-131 (p)	$7 \pm 5\times10^{-5}$	0.69 ± 0.12
I-133 (p)	$1.6 \pm 3\times10^{-5}$	0.5 ± 0.2
Cs-134 (p)	$2.8 \pm 0.6\times10^{-5}$	0.51 ± 0.07
Cs-137 (p)	$3.4 \pm 0.9\times10^{-5}$	0.59 ± 0.08

After each puff step (60 seconds), time-integrated air concentration (TIAC) is obtained by calculating concentration in one time-step and taking summation throughout simulation time. Plumes of particles released in the first hour are followed for the next 5 hours. Concentration is calculated using kernel approximation method for surface layer (0-5 m layer). Dose at gridpoints are then calculated using dose conversion factors from Lam et.al. [8].

Meteorological data for worst-case condition is obtained from site feasibility study monitoring program performed in the 1994-1995 period[11]. Frequency analysis was done to determine worst-case example in which there is high frequency of wind-direction consistently blowing to the land and populated area as well as calm or low wind-speed condition. A 12-hour duration of consistent wind direction from NW to N sectors in November 14, 1994 was identified. Five rain conditions were simulated: no rain, 5, 10, 15, and 20 mm/hr rain for the whole duration of simulation. These values are selected because rain analysis data shows that rain occured in 10 mm/hour or less almost 90% of the time. Larger values are added to see the effect of larger rates to the dose distribution and peak. In rain condition, at the end of every puff integration step, probability deposition of each particle is calculated and deposited particle height will be set to terrain height or at surface level.

During the worst-case situation, the Pasquill stability class varies between D and E (neutral and slightly stable).

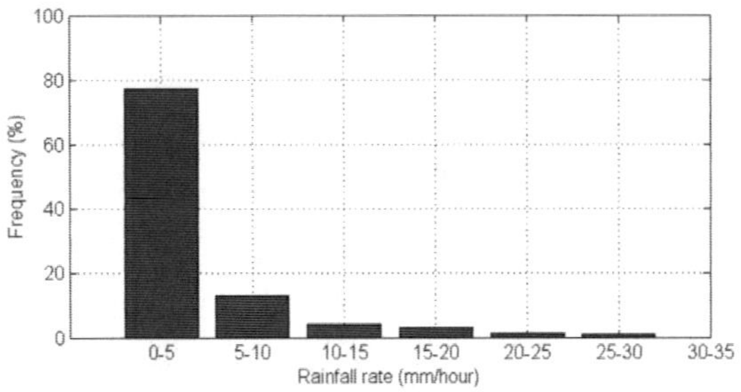

Figure 1. Rainfall rate in Ujung Lemahabang

3. Methodology

MODAL will be used to calculate external gamma exposure from I-131, Cs-137, Kr-88 and Xe-133 isotopes emitted through a 50 meter high stack in 6-hour duration. The modeled area is 100 km × 100 km with Cartesian grid size of $\Delta x = 1$ km, $\Delta y = 1$ km and $\Delta z = 0.05$ km. Total height of the model is 1500 meters. The air is incompressible and has uniform density anywhere in the grid.

Simulations are performed using isotopes of Cs-137, I-131, Kr-88, and Xe-133 each with a total strength of 0.011, 15.04, 62.000, and 8.37 million Curies respectively[7]. The plume is released from a 50 m high stack at a constant rate from second 8.500 to 30.000 after the actual accident happens. Two scenarios are prepared each containing different amount of Xe-133 release provided in **Table 1**. For each isotope, maximum value is taken from available data for conservative assumption.

A number of simulations are performed representing dry and rain condition with 5, 10, 15 and 20 mm/hr rain intensities. Worst-case meteorological condition is picked from meteorological database from on-site monitoring program performed during site feasibility study in 1994-1995 periods. Worst-case condition in this paper is defined as condition where wind is blowing to a consistent sector during the entire duration of radioactive release. This type of atmospheric condition will tend to produce higher TIAC for locations in the downwind direction from the source emission. Frequency analysis is performed to pick a period in the database where wind blow consistency is found.

4. Results and discussion

Simulations performed using TMI-2 source-term and worst-case meteorological data produce the peak doses for low-case and high-case scenarios are provided in **Table 3**.

Table 3. Peak dose

	Peak dose (mSv)	
	Low-case	High-case
Dry condition	1.668	6.892
Rain (5 mm/hr)	1.687	7.914
Rain (10 mm/hr)	1.861	7.621
Rain (15 mm/hr)	1.600	6.828
Rain (20 mm/hr)	1.705	7.798

As a comparison, the average peak dose from field monitoring after the accident in areas around TMI-2 was found to be around 0.2-1.1 mSv in different literatures. Due to the sparse location of monitoring stations, the measurement result may not resolve the actual peak dose in the area. Nevertheless, the measured peak dose is within one order of magnitude with the result of the simulation although the low-case source-term provide better representation to the actual source-term emitted during the accident. Peak dose gridpoint is located about 2.5 km from the point of emission. From the simulation, there is no apparent increase in maximum dose due to rain.

IOP Conf. Series: Journal of Physics: Conf. Series **799** (2017) 012015 doi:10.1088/1742-6596/799/1/012015

Dose distributions are provided in **Figure 3** and **4**, representing the dry condition and wet condition with 10 mm/hour rain respectively. High-dose areas are located a few kilometers away from the release point because the radioactive plume is released from a 50 m stack. The difference in dispersion pattern between dry and wet condition is not very noticeable in high dose area but is more visible in low dose area.

When doses at grid points are compared, slightly noticeable shift can be identified as described for 5 mm and 10 mm rains as given in Figure 5 and 6. In the event of rain, doses are generally higher at majority of grid points, especially at lower dose values below 0.15 mSv as indicated by the values of offsets and multipliers of the linear fit. Similar increases in dose are also observed for 15 mm and 20 mm rains.

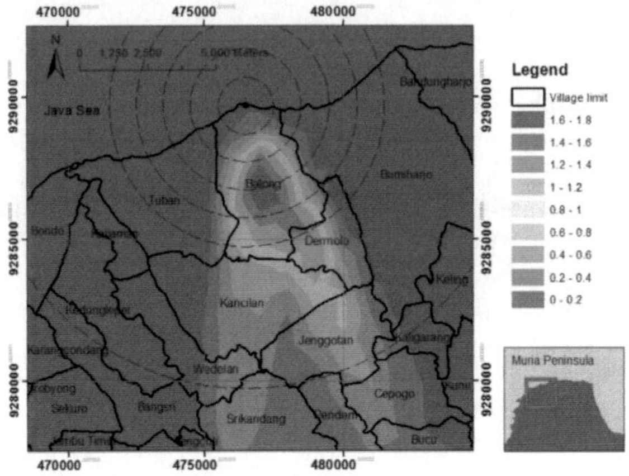

Figure 2. Dose Distribution (no wet deposition)

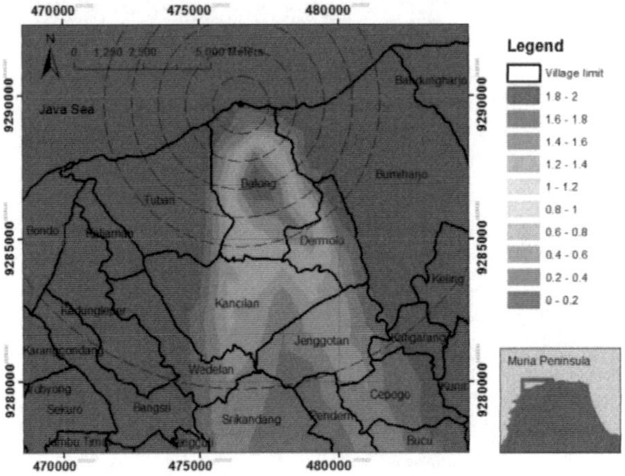

Figure 3. Dose Distribution (with wet deposition, 10 mm/hour rain)

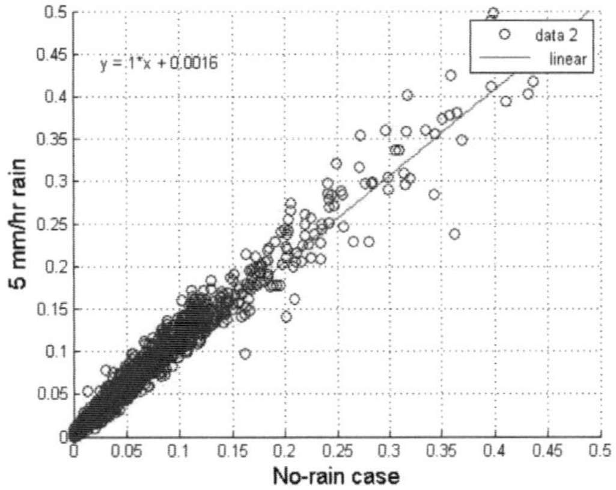

Figure 4. Grid point dose (0 and 5 mm rain)

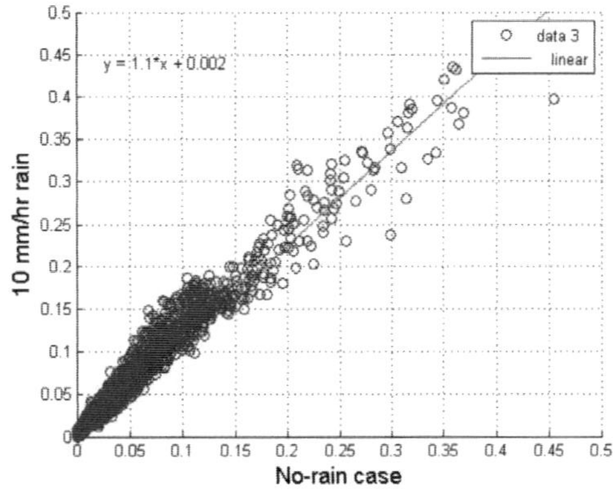

Figure 5. Grid point dose (0 and 10 mm rain)

5. Conclusion

In a consistent wind direction, dose accumulation is concentrated in areas downwind the release point. Fluctuation in wind direction may enhance the dispersion process and thus reduce the peak dose. Wind speed also induces more dispersion and will reduce peak dose but the location of the peak dose in this case will be father from the source.

Simulation using approximation for source-term and meteorological condition for TMI-2 reactor accident with assumed worst-condition for Muria peninsula has resulted in 1.668 mSv peak dose for low scenario and 6.892 mSv for the high scenario in dry condition. The peak dose is located in the radius between 3 to 5 kms in the direction of 12-hour consistent wind direction covering the middle part of Balong village. In the presence of rain during the simulation period has increased the dose at grid points but does not alter the distribution pattern significantly. At the dominant rainfall rate level of 5 mm/hr and 10 mm/hr for Ujung Lemahabang area, the increase in dose from wet deposition is considered to provide minor effect. In this simulation, the peak external dose is well below the regulatory limit of 50 mSv/year set by the Indonesian nuclear regulatory agency (BAPETEN) for equivalent dose for skin for general public.

Acknowledgment

The authors acknowledge the Ministry of Research and Technology and Nuclear Physics Laboratory of Institute Teknologi Bandung for supporting this research.

References

[1] Diehl S R, et.al. 1982 *Journal of Applied Meteorology* **21** 69.

[2] Uliasz M 1994 *Lagrangian Particle Dispersion Modeling in Mesoscale Applications. Environmental Modelling* II. ed. P. Zannetti (Southampton, Boston: Computational Mechanics Publication p 71)

[3] Brown M J Pal Arya S and William H Snyder 1993 *Journal of Applied Meteorology* **32** 490.

[4] Huang C H 1999 *Journal of Applied Meteorology* **38** 250.

[5] Haste T 2006 *Nuclear Engineering and Design* **236** 1099.

[6] Gudiksen P H and Dickerson M H 1990 *A Review Of Source-Term And Dose Estimation For The TMI-2 Rector Accident* (Lawrence Livermore National Laboratory).

[7] McColl N P and Prosser S L 2002 *Emergency Data Handbook NRPB-W19* (National Radiological Protection Board).

[8] Lam H Y Li S W and Yeung P Y 2005 *A Random-Walk Particle Dispersion Model for Radiological Accident Consequence Assessment* (Hong Kong Observatory).

[9] Ermak D R, Nasstrom J S and Taylor A G 1995 *Implementation of Random Displacement Method in the ADPIC Model Framework* (Lawrence Livermore National Laboratory).

[10] Sherman C A 1977 *Journal of Applied Meteorology* **17** 312.

[11] Sportisse B 2007 *Atmospheric Environment* **41** 2683.

[12] Molnar Jr. F, Szakaly T, Meszaros R and Lagzi L 2010 *Computer Physics Communication* **181** 105.

[13] Newjec 1996 *Topical Report on Meteorology, Rev. 3* (Jakarta: Newjec).

[14] http://www.nrc.gov/reading-rm/doc-collections/fact-sheets/3mile-isle.html

IOP Conf. Series: Journal of Physics: Conf. Series **799** (2017) 012016 doi:10.1088/1742-6596/799/1/012016

Yttria-Stabilized Zirconia Ceramic Deposition on SS430 Ferritic Steel Grown by PLD - Pulsed Laser Deposition Method

Abu Khalid Rivai[1,*], Mardiyanto[1], Agusutrisno[2], Edi Suharyadi[2]

[1]Center For Science and Technology of Advanced Materials – National Nuclear Energy Agency of Indonesia BATAN, Kawasan PUSPIPTEK Gd. 71, Tangerang Selatan, Banten, 15314, Indonesia

[2]Dept. of Physics, Faculty of Mathematics and Natural Science, Gadjah Mada University, Bulaksumur, Daerah Istimewa Yogyakarta 55281, Indonesia

Email: rivai.abukhalid@batan.go.id

Abstract. Development of high temperature materials are one of the key issues for the deployment of advanced nuclear reactors due to higher temperature operation. One of the candidate materials for that purpose is ceramic-coated ferritic steel that one of the functions is to be a thermal barrier coating (TBC). Thin films of YSZ (Ytrria-Stabilized Zirconia) ceramic have been deposited on a SS430 ferritic steel using Pulsed Laser Deposition (PLD) at Center For Science and Technology of Advanced Materials laboratory – National Nuclear Energy Agency of Indonesia (BATAN). The thin film was deposited with the chamber pressure range of 200-225 mTorr, the substrate temperature of 800°C, and the number of laser shots of 3×10^4, 6×10^4 and 9×10^4. Afterward, the samples were analyzed using Scanning Electron Microscope – Energy Dispersive X-ray Spectroscope (SEM-EDS), X-Ray Diffractometer (XRD), Atomic Force Microscope (AFM) and Vickers hardness tester. The results showed that the YSZ could homogeneously and sticky deposited on the surface of the ferritic steel. The surfaces were very smoothly formed with the surface roughness was in the range of 70 nm. Furthermore, thickness, composition of Zr^{4+} dan Y^{3+}, the crystallinity, and hardness property was increased with the increasing the number of the shots.

1. Introduction

Development of high temperature materials are one of the key issues for the deployment of systems in many fields such as advanced nuclear reactors due to the need of higher temperature operation. One of the candidate materials for that purpose is ceramic coating on the structural materials [1-5]. The coating is able to protect the substrate materials from corrosion [6]. The techniques of the coating are mostly done by electron beam physical vapor deposition (EB-PVD) [2] and plasma spray coating [4,5,7]. Nevertheless, in order to minimize micro porous the new technique for coating i.e. pulsed laser deposition (PLD) has been reported [1]. It has been reported that the PLD technique is used in many field [8-15] and able to deposit not only ceramic but also polymer material [16].

In this study ceramic-coated ferritic steel that one of the functions is to be a thermal barrier coating (TBC) is reported. YSZ (Ytrria-Stabilized Zirconia, Y_2O_3 – ZrO_2) ceramic was used as the target for coating using PLD technique. YSZ material has advantages as coating materials to protect the substrate at high temperature [17]. The purpose of the study is to develop high quality of coating layer with smooth roughness of surface and minimizing porosity.

2. Experiment

Thin films of YSZ (Ytrria-Stabilized Zirconia) ceramic have been deposited on a SS430 ferritic steel using Pulsed Laser Deposition (PLD) at laboratory facilities of Center For Science and Technology of Advanced Materials –National Nuclear Energy Agency of Indonesia (BATAN). Figure 1 (a) shows the schematic of the PLD and Figure 1 (b) shows the plume formation during deposition. It can be seen this PLD is able to use multiple target in the carrousel. Furthermore, performance of the plume during deposition is able to be seen and analyzed during deposition through a glass window. The parameter condition of the experiment can be seen in the Table 1. The target was 8YSZ (8%mol Y_2O_3 – 92%mol ZrO_2) and the substrate was SS430 ferritic steel with the content of chromium is up to 15%. Solid state laser of Nd:YAG was used in this PLD with the wavelength of 266 nm and the energy of ~100 mJ. Thin film was deposited with the constant oxygen flow injection of 40 sccm (Standard Cubic Centimeters per Minute) that produce a chamber pressure with the range of 200-225 mTorr. The substrate temperature during deposition was of 800°C, and the numbers of laser shots were 3×10^4, 6×10^4 and 9×10^4.

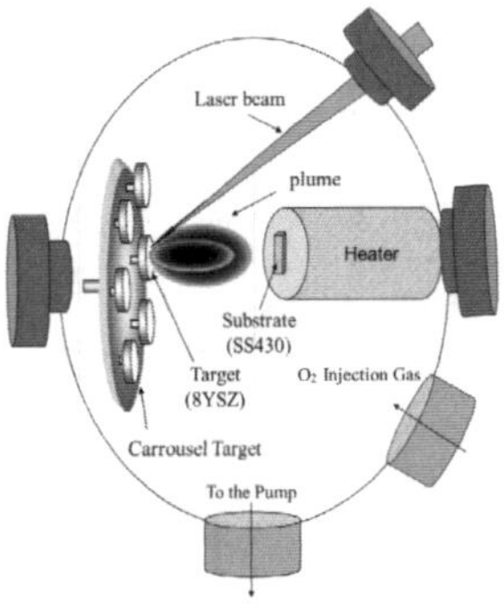

(a) (b)

Figure 1. PLD – Pulsed Laser Deposition (a) schematic (b) plume formation during depostion.

Table 1. Parameter of experiment

Parameter	Condition
Target material	8YSZ (8%mol Y_2O_3 – 92%mol ZrO_2)
Substrate material	Steel (SS430)
Temperature of substrate	800 °C
Number of shots	$3x10^4$, $6x10^4$ and $9x10^4$
Chamber pressure	200-225 mTorr
Gas background	Oxygen
Laser wavelength	266 nm
Laser energy	~100 mJ
Laser frequency	10 Hz
Laser repetition pulsed	5 ns

IOP Conf. Series: Journal of Physics: Conf. Series **799** (2017) 012016 doi:10.1088/1742-6596/799/1/012016

After deposition of the steel with YSZ target, the samples were analyzed using Scanning Electron Microscope – Energy Dispersive X-ray Spectroscope (SEM-EDS), X-Ray Diffractometer (XRD), Atomic Force Microscope (AFM) and Vickers hardness tester.

3. Results and Discussion

Figure 2 show SEM micrographs of samples cross section to investigate the film deposition on the steel with various shots i.e. $3x10^4$ shots, $6x10^4$ shots and $9x10^4$ shots. It was found that the thickness of the coating layer were ~0.5, 1, and 1.1 μm for $3x10^4$ shots, $6x10^4$ shots, and $9x10^4$ shots, respectively. The results showed that the thickness of the coating layer was increase with the increasing of the shots.

Figure 3 show SEM micrographs of samples surface with EDS area analyses to investigate the deposition of YSZ on the steel with various shots i.e. $3x10^4$ shots and $9x10^4$ shots. The EDS analyses of zirconium and yttrium percentages on the steel with various shots are shown in Table 2. The results showed that the percentages of zirconium and yttrium were increased with the increasing of the shots. The percentages analyses mean that the thickness of the layer was increased with the increasing of the shots.

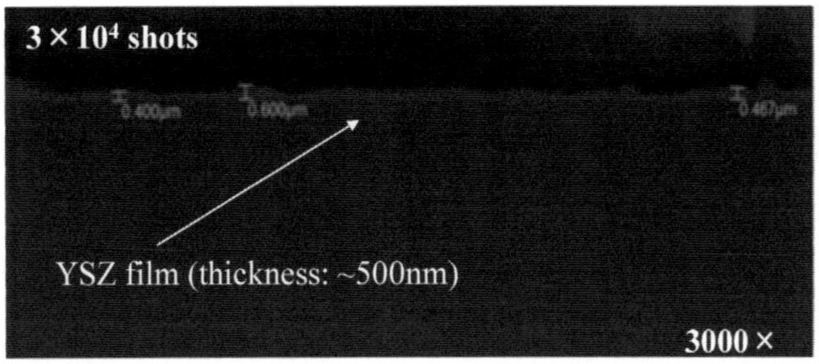

(a)

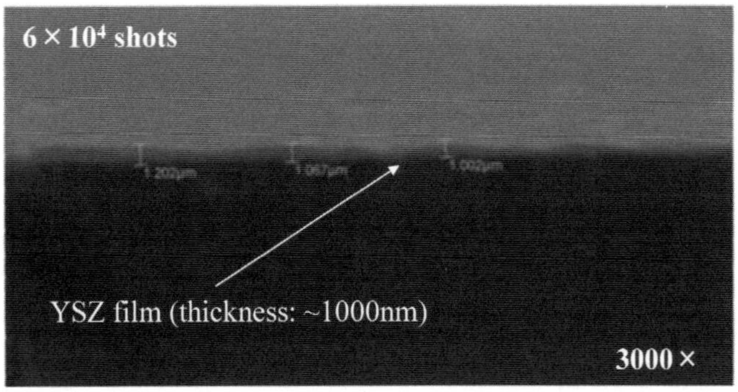

(b)

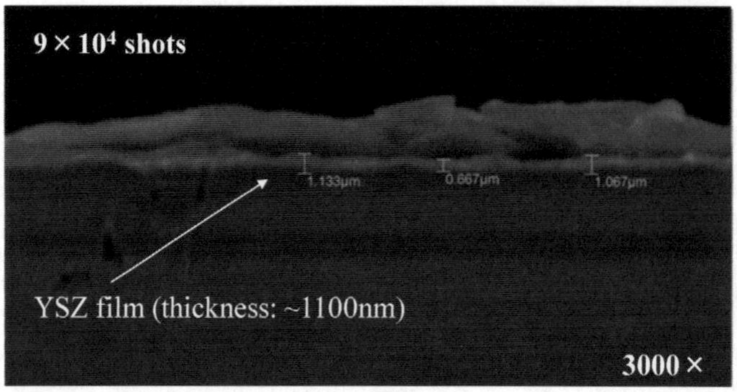

(c)

Figure 2. SEM micrographs of samples cross section to investigate the film deposition on the steel with various shots (a) $3x10^4$ shots (b) $6x10^4$ shots and (c) $9x10^4$ shots

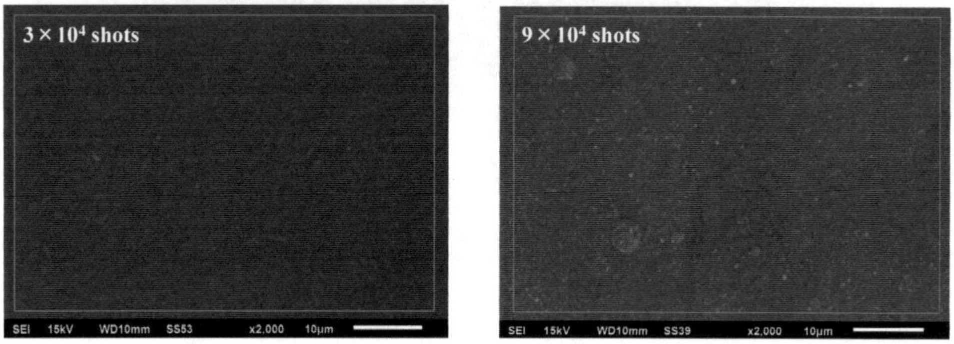

Figure 3. SEM micrographs of samples surface with EDS area analyses to investigate the deposition of YSZ on the steel with various shots (a) $3x10^4$ shots and (b) $9x10^4$ shots

Table 2. EDS analyses of zirconium and yttrium percentages on the steel with various shots

Elements /	$3x10^4$ shots		$9x10^4$ shots	
Parameter	% mass	% atom	% mass	% atom
Zr	8.33	3.72	34.25	13.85
Y	2.88	1.33	6.74	2.87

Figure 4 show the AFM - Atomic Force Microscope topography analyses of the samples's surface (2 and 3 dimensions) for $3x10^4$ number of shots, $6x10^4$ number of shots and $9x10^4$ number of shots. The results showed that the roughness of the coating layer was in the nano-meter scale i.e. ~70 nm for all the cases. It is found that the PLD technique for ceramic coating could develop very smooth layer and minimize the formation of micro porous. As comparison analysis, it has been reported that using plasma sprayed technique the surface roughness was 10 μm and using EB-PVD technique the surface roughness was 1 μm [18]. Therefore, PLD technique has advantage to develop very smooth surface of coating.

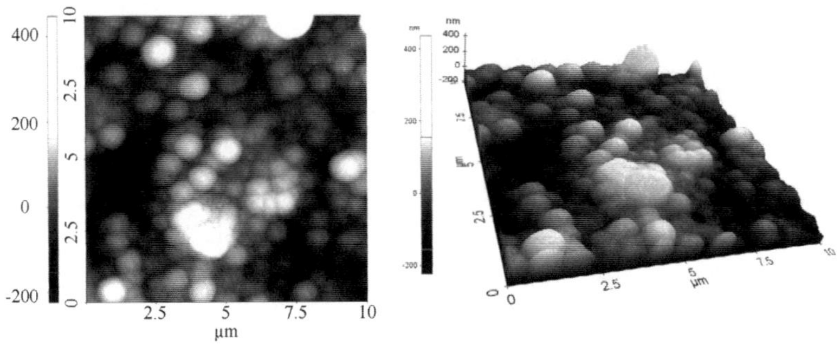

(a)

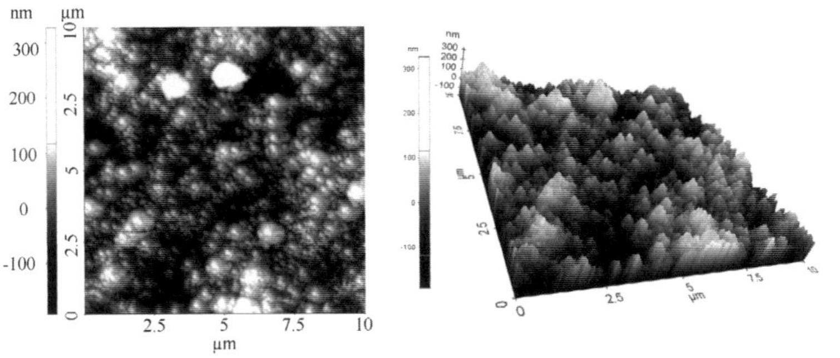

(b)

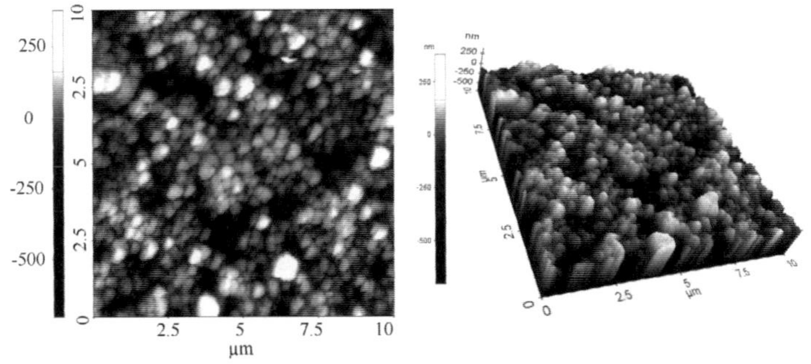

(c)

Figure 4. AFM - Atomic Force Microscope topography analyses of the sample's surface (2 and 3 dimensions) (a) $3x10^4$ number of shots (b) $6x10^4$ number of shots (c) $9x10^4$ number of shots

Figure 5 show XRD analyses of the samples for 2 between 20-90° and 2 between 20-40°. The results showed that the ceramic YSZ was deposited on the layer of the substrate with the indication of YSZ peaks. Furthermore, the substrate peaks were decrease with the increasing of the PLD shots and the crystanillity of the YSZ layer was increase with the increasing of the shots. It happen because with the increasing of the shots the thickness of the layer was increase.

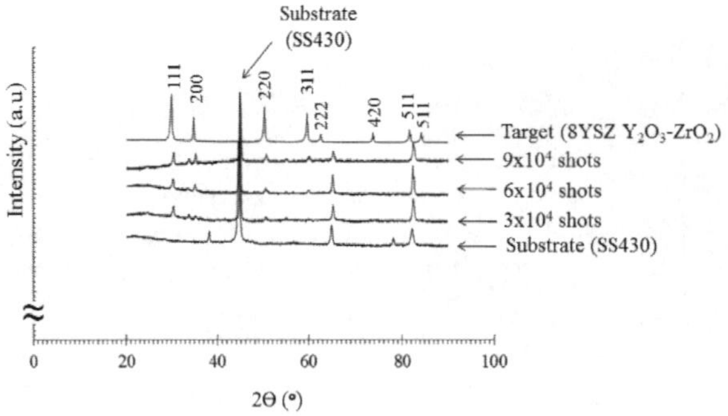

(a)

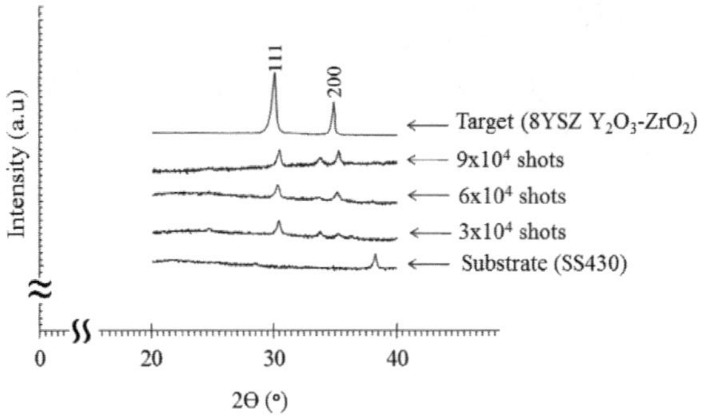

(b)

Figure 5. XRD analyses of the samples: (a) 2 between 20-90°, and (b) 2 between 20-40°

Table 3. Surface hardness of the samples

Number of shots	Hardness (VHN)
$3x10^4$	167
$6x10^4$	182
$9x10^4$	184

Table 3 show the surface hardness of the samples with the various of the number of shots. The results showed that the hardness property was increased with the increasing the number of the shots. The surface hardness was contributed from the hardness of the thin film and the hardness of the substrate. The increasing of the number of shots increased the thickness of the coating layer therefore it minimized the contribution of the hardness of the substrate. It means that the hardness of YSZ is higher than the steel and finally the hardness is only the property of the YSZ coating layer.

IOP Conf. Series: Journal of Physics: Conf. Series **799** (2017) 012016 doi:10.1088/1742-6596/799/1/012016

4. Conslusions

Thin films of YSZ (Ytrria-Stabilized Zirconia) ceramic have been deposited on a SS430 ferritic steel using Pulsed Laser Deposition (PLD) technique with the chamber pressure range of 200-225 mTorr, the substrate temperature of 800°C, and the number of laser shots of 3×10^4, 6×10^4 and 9×10^4. The surfaces were very smoothly formed with the surface roughness was in the range of 70 nm. Furthermore, thickness, composition of Zr^{4+} dan Y^{3+}, the crystallinity, and hardness property was increased with the increasing the number of the shots. The PLD technique is potential to be used for coating of YSZ on the ferritic steel for the development of high temperature materials.

Acknowledgment

Authors would like to express gratitude to Drs. Bambang Sugeng and Dr. Arbi Dimyati for their help in characterization of the samples using XRD and SEM-EDS, respectively, and Head of Center For Science and Technology of Advanced Materials (PSTBM) – BATAN for his support and encouragement. This research is supported by Indonesian Government through DIPA PSTBM-BATAN.

References

[1] Sujaya C, Shashikala H D, Umesh G, and Hegde A C 2011 *J.Electrochem. Sci. Eng.* **1** 1.
[2] Movchan B A and K. Yu Yakovchuk 2014 *J. Coating Sci. Tech.* **1** 96.
[3] Dharam K 2015 *Intl. J. Innovative Research in Sci. Eng. Tech.* **4** 2811.
[4] Mohamed S M, Soha A A, Madiha. A S, and Khalid F A 2012 *Int. J. Electrochem. Sci.* **7** 2811.
[5] Nicholas C, Kent V E, Todd S, and Nicolaie M 2014 *Coatings* **4** 630.
[6] Ibtihal A M, Walaa W J, and Lubna A K 2013 *Intl. J. Research in Eng. & Tech.* **1** 77.
[7] Armelle V, Christian M, Nickolas J. T, and Christophe C 2014 *Plasma Chem Plasma Process* DOI 10.1007/s11090-014-9600-y.
[8] Dikovska A Og, Atanasova G B, Avdeev G V, and Strijkova V Y 2016 *J. of Physics: Conference Series* **700** 012024.
[9] Sposito A, May-Smith T C, Stenning G B G, de Groot P A J, and Eason R W *Optical Materials Express* **3** 5.
[10] Juguang H, Huabin T, Xiaodong L, Zhongkuan L, Huiqun C, Qiwen L, Yi L, Jinghua L, and Pei W 2012 *Intl J. Photoenergy* **2012** 1.
[11] Cancea V N, Birjega R, Ion V, Filipescu M, and Dinescu M 2012 *Physics AUC* **22** 50.
[12] Agusutrisno, Edi S, Mardiyanto, Abu K R 2015 *Spektra: Jurnal Fisika dan Aplikasinya* **16** 34.
[13] Mardiyanto, Abu K R, Agusutrisno, Edi S 2015 *Prosiding Seminar Nasional Teknologi Energi Nuklir* **ISSN: 2355-7524** 208.
[14] Agusutrisno 2015 *Master Thesis* Dept. Physics Gadjah Mada University.
[15] Dave H A B, Matthijn D, and Guus R 2014 *J. Phys. D: Appl. Phys.* **47** 034006.
[16] Caricato A P, Luches A, Leggieri G, Martino M, and Rella R 2012 *Vacuum* **86** 661.
[17] Hengbei Z, Fengling Y, Ted D B, Haydn N G W 2006 *Acta Materialia* **54** 5195.
[18] Nicholls J R, Lawson K J, Johnstone A, and Rickerby D S 2002 *Surface and Coatings Techn.* **151-152** 383.

IOP Conf. Series: Journal of Physics: Conf. Series **799** (2017) 012017 doi:10.1088/1742-6596/799/1/012017

Analysis of High Temperature Reactor Control Rod Worth for the Initial and Full Core

Hammam Oktajianto[1], Evi Setiawati[1], Khoirul Anam[1], Heri Sugito[1]

[1]University of Diponegoro, Semarang, Indonesia

Email: hammam_oktajianto@st.fisika.undip.ac.id

Abstract. Control rod is one important component in a nuclear reactor. In nuclear reactor operations the control rod functions to shut down the reactor. This research analyses ten control rods worth of HTR (High Temperature Reactor) at initial and full core. The HTR in this research adopts HTR-10 China and HTR- of pebble bed. Core calculations are performed by using MCNPX code after modelling the entire parts of core in condition of ten control rods fully withdrawn, all control rods in with 20 cm ranges of depth and the use of one control rod. Pebble bed and moderator balls are distributed in the core zone using a Body Centred Cubic (BCC) lattice by ratio of 57:43. The research results are obtained that the use of one control rod will decrease the reactor criticality of 2.04±0.12 %Δk/k at initial core and 1.57±0.10 %Δk/k at full core. The deeper control rods are in, the lesser criticality of reactor is with reactivity of ten control rods of 16.41±0.11 %Δk/k at initial core and 15.43±0.11 %Δk/k at full core. The results show that the use of ten control rods at full core will keep achieving subcritical condition even though the reactivity is smaller than reactivity at initial core.

1. Introduction

Attention scientist of reactor technology and nuclear energy in the world against the High Temperature Reactor (HTR) has increased in the past decade. Inherent safety characteristics and capabilities to produce energy economically are the main factors that attract many people to study and develop HTR. The HTR utilizes graphite as the moderator at the same reflector and the fuel is a spherical particle (pebble-bed) with UO_2 composition as a neutron generator [1].

HTR is the types of gas-cooled high temperature reactor. HTR core design in this study is a blend HTR 10 in China with HTR pebble-bed. Thermal reactor power is 10 MW with inlet and outlet helium temperatures of 250°C and 700°C. HTR design is a cylindrical with helium gas as a coolant and graphite as a moderator. In addition, the HTR uses pebble-bed fuel composed a large amount of particles of TRISO in graphite metrics. The TRISO particle is coated fuel particles by a radius of 175-300 μm. According to Hammam, kernel radius of 225 μm with enrichment of 16% can use in HTR due to reactor has been critical [2-5].

The criticality of reactor is a variable that describes the state of the reactor in order to operate optimally. The reactivity of reactor can be controlled by control rod. Therefore, it is needed to determine the control rods worth so that reactor is in critical or subcritical condition at initial and full core [6].

Calculation of the reactor criticality at any control rods depth variations is done with modelling techniques of MCNPX which is a software that can simulate the interaction of particles in a reactor with a Monte Carlo approach. The modelling is done by making the geometry of reactor, control rod and pebble-bed which is distributed by using body-centred cubic lattice in the reactor core.

2. Experimental Method

Materials used in this research were HTR10, HTR pebble bed database and continuous energy nuclear data library ENDF/B-VII. Modelling of HTR uses the Monte Carlo code MCNPX. MCNP (Monte Carlo N-particle) is a general-purpose, continuous-energy, generalized-geometry, time-dependent, coupled neutron, photon and electron Monte Carlo transport code. MCNP is also capable of calculating the multiplication factor (criticality) of fissile systems. A system is defined by generating cells bounded by surfaces in three dimensions. Any kind of geometry can be defined as a cell and this cell can be rotated and moved to anywhere in the space. MCNP can simulate particle transport. Monte Carlo simulates individual particles and recording some aspects (tallies) of their average behaviour [7].

Criticality evaluations are done based on the principle of neutron balance. The number of neutrons in each generation is taken into account and comparison is made with the number of neutrons in the consequent generation. All possible mechanisms for the birth and loss of neutrons are accounted in bookkeeping. Thus, effective multiplication factor is evaluated for a given cycle. Each fission neutron is generated randomly out of possible locations containing fissile material. In order to generate statistical basis, simulations are repeated as many times as desired.

The initial step is to calculate atom density from reactor and fuel, then to model the reactor core with diameter of 180 cm and height of 197 cm. Reactor core is surrounded by a graphite reflector, while the graphite reflector is surrounded by layer of boronated carbon bricks and core is filled by pebble bed and moderator balls. On the side reflectors near the active core there are ten boreholes with 130 mm diameter for the insertion of control rods and three boreholes of 130 mm diameter for irradiation. On the side of the reflector there are twenty flow channels in the form borehole with 80 mm diameter for helium inlet. Pebble-bed is distributed in the reactor core which is composed of many triso particles. Triso is a fuel that is composed of Uranium Dioxide coated by four outer layers: Carbon, IpyC (Inner Pyrolytic Coating), SiC (Silicon Carbides) and OpyC (Outer Pyrolytic Coating). Both of PyC have density differences. The first layer next to kernel (UO_2) has lower density than others that is functioning to intercept gas of fissile product. SiC layer serves as a barrier to the production of fissile active movement like Cs, Sr and Ag. SiC is also a mechanical and chemical barrier at high temperatures [8]. The basic characteristics of the fuel elements are shown in Table 1.

Table 1.Fuel element characteristics

Fuel kernel	
Diameter of ball (cm)	6.0
Diameter of fuelled region (cm)	5.0
Density of graphite in matrix and outer shell (g/cm^3)	1.73
Enrichment of 235U (w%)	16
Equivalent natural boron content of impurities in uranium (ppm)	4.0
Equivalent natural boron content of impurities in graphite (ppm)	1.3
Radius of the kernel (μm)	225
UO_2 density (g/cm^3)	10.4

Coatings	
Coating layer materials (starting from kernel)	C/IPyC/SiC/OPyC
Coating layer thickness (mm)	0.09/0.04/0.035/0.04
Coating layer density (g/cm^3)	1.1/1.9/3.18/1.9

Pebble-bed and moderator balls is distributed in the core zone of the HTR using a body-centred cubic (BCC) lattice by packing fraction and the percentages of the pebble and moderator balls of 0.61 and 57:43. BBC lattice modelling in MCNP uses lattice option at any active core height variation. MCNP model for BBC lattice, pebble-bed and TRISO is shown in Figure 1 and 2, while MCNP reactor model is shown in Figure 3.

There are ten control rods placed in the side reflector of reactor. Boron Carbide (B4C) is used as the neutron absorber. Each control rod contains five B4C ring segments which are haused in the area between an inner and an outer sleeve of stainless steel. The inner and outer diameter of the B4C ring is 60 mm and 105 mm respectively, while the length of each ring segment is 487 mm. the inner/outer diameters of the inner and outer stainless steel sleeves are 55 mm/59 mm and 106 mm/110 mm, respectively. The length of each joint is 36 mm. the lengths of the lower and upper metalic end are 45 mm and 23 mm. The basic characteristics of the control rod are shown in Table 2. Figure 4 shown control rod of reactor.

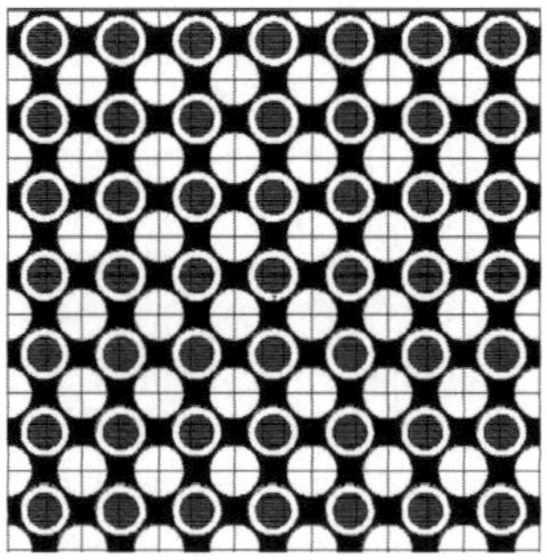

Figure 1. Pebble-bed and Moderator ball is in BCC lattice in reactor core

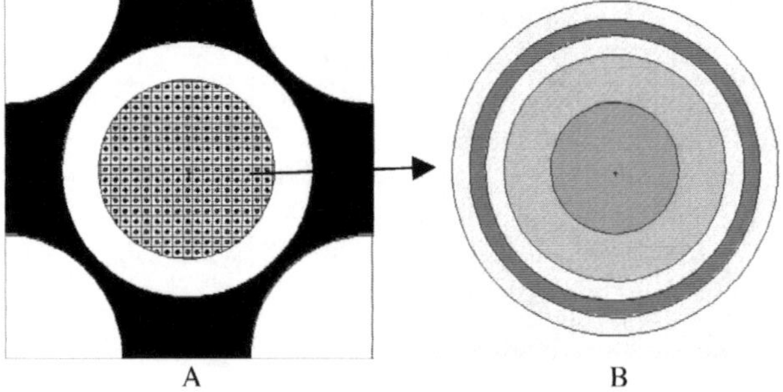

Figure 2. Pebble-bed with Moderator ball in every corner (A) and Triso coated by UO_2, Carbon, IPyC, SiC and OPyC (B).

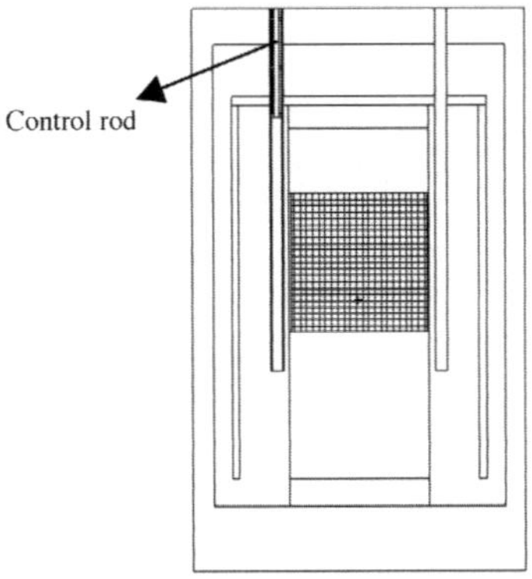

Figure 3. MCNP model of HTR geometry at XZ coordinate

Figure 4. MCNP model of control rod at XZ coordinate

Table 2. Control rod characteristics

Elemen	
B4C (g/cm3)	7.9
Cr (%)	18
Fe (%)	68.1
Ni (%)	10
Si (%)	1
Mn (%)	2
C (%)	0.1
Ti (%)	0.8
Description	
Control rod channel radius (cm)	6.5
Radial position of channel center (cm)	102.1
Length of B4C segment (cm)	48.7
Length of bottom metalic end (cm)	4.5
Length of metalic joins (cm)	3.6
Length of top metalic end (cm)	2.3
Inside radius of inner stainless-steel sleeve (cm)	2.75
Thickness of stainless-steel sleeve (cm)	0.2
Thickness of gap between sleeve and B4C (cm)	0.05
Thickness of B4C annulus (cm)	2.25
Density of B4C (cm)	264.7
Length of control rod (cm)	1.7

After reactor core has been modelled, the next process is calculation of reactor criticality done in initial and full core in each control rod condition: fully withdrawn, one control road fully in and all control rods in with 20 cm ranges of depth with the number of neutrons simulated in KCODE card and neutron source in SDEF card which are specified by reactor core design. 5000 neutrons in each cycle are simulated by estimation criticality value (K_{eff}) of 1.0 selected in order that the final accumulation results are expected nearly equal to the critical condition. Using skipping 10 cycles is done before data accumulation of criticality value from a total of 210 cycles to prevent convergence of the source and that the fission sources can be stable before criticality values are used to average its final estimation. SDEF card is utilized to specify fission source distributions in reactor core and S(α,β) graph.01t of thermal neutron scattering data is applied in all materials containing graphite to consider binding effect at thermal neutron and graphite moderator under energy of 4 ev.

3. Results and Discussion

In MCNP In the calculation of HTR pebble-bed, pebble-bed core model is approximated by utilizing a BCC lattice. Repeating structure of MCNP leads to the emergence of partial pebble around the core which can add extra fuel into the core. Excess fuel contributed by this partial pebble is eliminated by reducing the volume of the core where a pebble packing fraction is maintained unchanged. This approach relies on an exclusion zone which compensates contribution of partial pebble. The size of the exclusion zone is given by the pebble radius determined by the ratio of the number of fuel pebble with the number of pebble in unit cells, thus obtained exclusion zone thick is 1.71 cm around the core with fuel pebble and moderator ratio is 57:43.

All of MCNP5 calculations utilize the continuous energy nuclear data library ENDF/B-VII with a temperature of 27°C. From the MCNPX calculations, the results are obtained that the use of one control rod will decrease the reactor criticality of 2.04±0.12 %Δk/k at initial core and 1.57±0.10 %Δk/k at full core. The results of HTR criticality calculation with ten control rods in of 20 cm ranges are plotted in graph and are shown in Figure 5 and the control rods worth are shown in Figure 6.

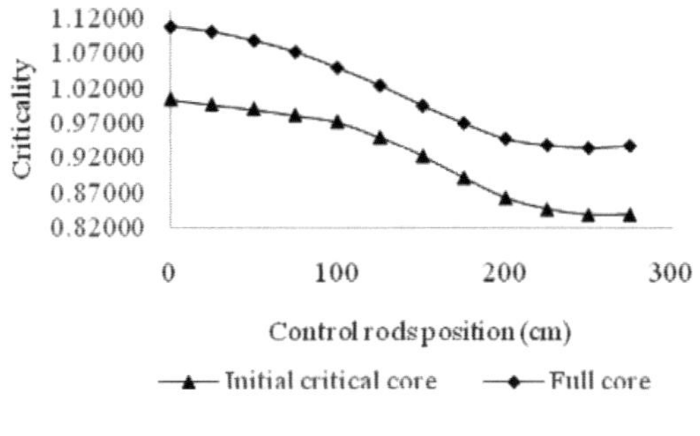

Figure 5. The results of criticality reactor (K_{eff}) in each control rods position

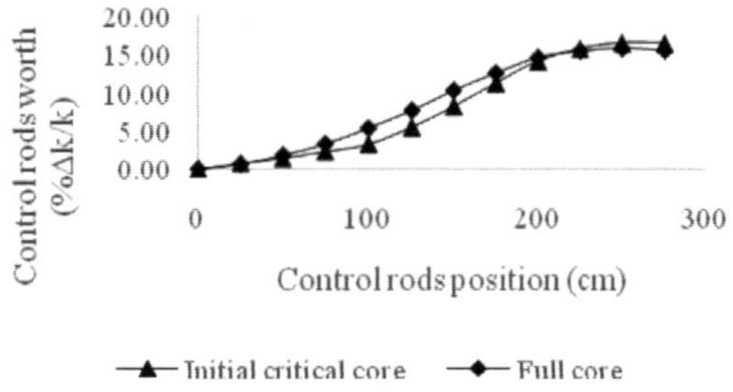

Figure 6. The control rods worth in each control rods position

IOP Conf. Series: Journal of Physics: Conf. Series **799** (2017) 012017 doi:10.1088/1742-6596/799/1/012017

From the results shown in Figure 5 and 6 seem that the deeper control rods are in, the lesser criticality of reactor is with reactivity of 16.41±0.11 %Δk/k at initial core and 15.43±0.11 %Δk/k at full core. That occur as the use of B4C in control rod where B_4C has a function as the neutron absorber. The Boron material is a effective material to absorb neutron. When many B_4C are in core, many neutrons will be absorbed. That result shown that control rod can make subcritical condition either in initial core or in full core.

4. Conclusion

From the results explained above can be concluded that from the results of MCNPX calculations the use of control rods can make reactor to be subcritical where the number of neutrons decrease. The deeper control rods are in, the lesser criticality of reactor is. This is occurred due to there are many Boron in B_4C used in control rods so activity of fission in reactor decreases. However, the use of ten control rods at full core still achieves subcritical condition even though reactivity smaller than reactivity at initial core.

References

[1] Holbrook 2008 NRC Licensing Strategy Development for the NGNP (Washington DC)
[2] IAEA 2003 Evaluation of High Temperature Gas Cooled Reactor Performance: Benchmark Analysis Related to Initial Testing of The HTTR and HTR-10
[3] G. Hosking and T.D. Newton 2007 OECD/NEA Nuclear Science Committee Results of Benchmark Considering A High-Temperature Reactor (HTR) Fuelled With Reactor-Grade Plutonium. Physics of Plutonium Recycling; Volume VIII
[4] Zuhair 2012 Investigasi Kritikalitas HTR (High Temperature Reactor) Pebble Bed Sebagai Fungsi Radius dan Pengayaan Bahan Bakar Kernel. Indonesia Journal of Applied Physics, vol.2. pp. 146
[5] O. Hammam, S. Evi and R. Very 2014 Modelling of HTR (High Temperature Reactor) Pebble-Bed 10 MW to Determine Criticality as A Variations of Enrichment and Radius of the Fuel (Kernel) With the Monte Carlo Code MCNP4C, International Journal of Science and Engineering, 8(1),42-46
[6] S. Volkan and Ç. Üner 2002 HTR-10 full core first criticality analysis with MCNP Nuclear Engineering and Design 222 (2003) 263–270
[7] J. F. Briesmeister 1992, Los Alomos National Laboratory MCNP-A General Monte Carlo N-Particle Transport Code
[8] R. Didiek Herhady and R. Sukarsono 2007 Pengaruh Suhu dan Waktu Sintering Terhadap Kualitas Bahan Bakar Kernel UO_2 Dalam Furnace Jenis Fluiduzed Bed, Prosiding Seminar Nasional ke-13 Teknologi dan Keselamatan PLTN Serta Fasilitas Nuklir

IOP Conf. Series: Journal of Physics: Conf. Series **799** (2017) 012018 doi:10.1088/1742-6596/799/1/012018

Synthesis Oxide Dispersion Strengthening Stainless Steel doped with Nano Zirconia by Mechanical Alloying

Pawawoi[1], Irfan Widiansyah[1], and Djoko Hadi Prajitno[2]
[1] Metallurgy Engineering, UNJANI Jl. Gatot Subroto Bandung, Indonesia
[2]PSTNT-BATAN Jl. Tamansari 71 Bandung 40132, Indonesia

Email: djokohp@batan.go.id

Abstract. The oxide dispersion strengthening stainless steel of Fe-11.5wt%Cr and Fe-11.5wt%Cr-1%ZrO_2 alloy by mechanical alloying method were synthesized by planetary ball milling. The methods employed for study were designing of Fe-11.5wt%Cr and Fe-11.5wt%Cr-1%ZrO_2 proportion of composition alloy which is plotted to *Schaffler* diagram to get ferritic/martensitic stainless steel. After MA the ODS powders were compaction with pressure 80kg/mm2 and followed by sintering at the temperature of 900,1000 and 1100° C under high purity argon atmosphere for 1 hour. Characterization by XRD is used to examination phase present. Optical microscopy and SEM is used to get image microstructures. XRD analysis resulting the ferritic and martensitic is a major and minor phase respectively. There are not significant differences in the microstructure between Fe-11.5wt%Cr and Fe-11.5wt%Cr-1wt%ZrO_2. An increase in the sintering temperature shift the microstructure from dendritic to equaxed. EDS examination showed that zirconia exit in the alloy Fe-11.5wt%Cr-1wt%ZrO_2.The addition of 1 % nano-zirconia (ZrO_2) into Fe-Cr alloy while milling process was resulted a higher Hardness Vickers Values rather than without zirconia addition. Average value of Hardness Vickers values was resulted 135.5 HV for Fe-11.5wt%Cr whereas 138.4 HV for Fe-11.5wt%Cr-1wt%ZrO_2.

1. Introduction

Current development of high-performance nuclear structural materials for breeding-blanket structural components, which will be exposed to high fluxes neutrons is one of the major challenges in future fusion reactors materials. The requirement of structural materials for the blanket components to a large degree dictates the design of the reactor systems. The selection of suitable structural materials is based on mechanical, thermophysical, corrosion and compatibility, low neutron inducedradioactivity, and resistance to radiation-induced damage phenomena like material hardening embrittlement [1].

Oxide-dispersion strengthened (ODS) stainless steels have been considered as promising high-temperature materials such as nuclear materials. High-chromium ferritic/martensitic steels are considered for the Generation IV nuclear reactors operated at elevated-temperature in-core applications. Basically, these steels contain 9–12% Cr to improve high temperature oxidation. [2,3] Mechanical alloying (MA) is one of the methods to synthesize material uniformly by disperse nano-sized oxide particles such as yttria(Y2O3) in the metal matrix. This technique can be also employed to fabricate nano Fe alloys with the oxide dispersion [4,5,6].

The purpose of this paper is to characterize the microstructural and mechanical properties (hardness) of ferritic/matensitic stainless steel of Fe-11.5wt%Cr and Fe-11.5wt%Cr-wt%ZrO2 as a candidate structural material for nuclear reactors application [7,8].

2. Experimental methods

The powder mixture with the nominal composition of Fe-11.5wt%Cr and Fe-11.5wt%Cr-1wt%ZrO2 alloy was mechanically alloyed using Planetary Ball Milling (PBM) equipped with stainless steel vial for 2 hours. The milling balls were made of Zirconia with diameter of 2 mm. Ball-to-powder weight ratio was 10:1. The speed of rotating arm was 580 r.p.m. After MA the ODS powders were compaction with pressure 80kg/mm2 and followed to sintering at the temperature of 900,1000 and 1100° C under high purity argon atmosphere for 1 hour. Micro hardness measurements were performed by using micro hardness tester HV 1000 equipped with a Vickers diamond pyramid and applying load of 0.98N for 15s. Each result is the average of at least 5 measurements. Characterization XRD by PHILIPS type PW 1835 device, using the Cu-K_ radiation with l=0.15406 nm. Optical microscopy and SEM is used to get image microstructures.

3. Result and discussions

X-ray diffraction patterns of the Fe-11.5wt%Cr and Fe-11.5wt%Cr-1wt%ZrO2 alloy after sintering from planetary ball mill are shown in Figs. 1a and b, respectively. The peaks of elemental powder ZrO2, was not appeared in the diffraction pattern. The phase present is dominated by ferritic phase and martensitic is minor phase.

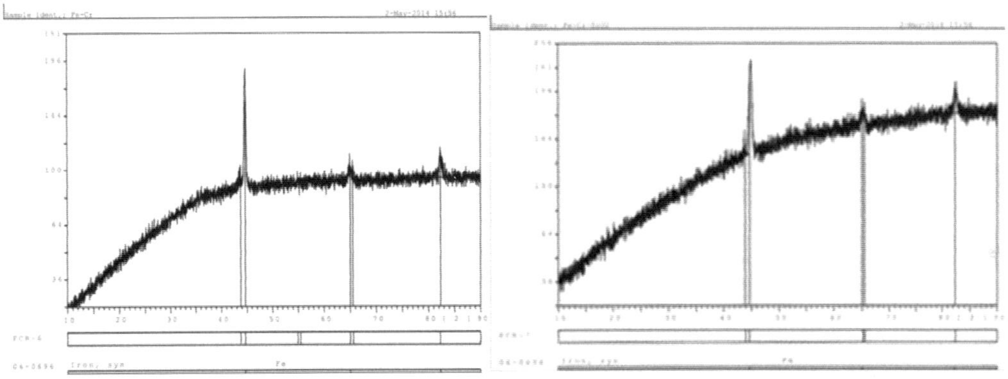

A b

Figure 1. a) Fe-11.5wt%Cr b) Fe-11.5wt%Cr-1wt%ZrO2 sinter at 11000C

Figure 2,3, and 4 shows representative optical and SEM-EDS micrographs of this Fe-11.5wt%Cr and Fe-11.5wt%Cr-1wt%ZrO_2 stainless steel after sintering at 900^0C,1000^0C and 1100^0C. According to optical microscopy, there are not significant differences in the microstructure between Fe-11.5wt%Cr and Fe-11.5wt%Cr-1wt%ZrO_2 After sintering at temperature 900^0C and 1000^0C. But an increase in the sintering temperature shift microstructure significantly at 1100^0C. The microstructure changed from dendritic to equated. Fig.4 showed that EDS examination depicted that zirconia exit in the alloy Fe-11.5wt%Cr-1wt%ZrO_2.

IOP Conf. Series: Journal of Physics: Conf. Series **799** (2017) 012018 doi:10.1088/1742-6596/799/1/012018

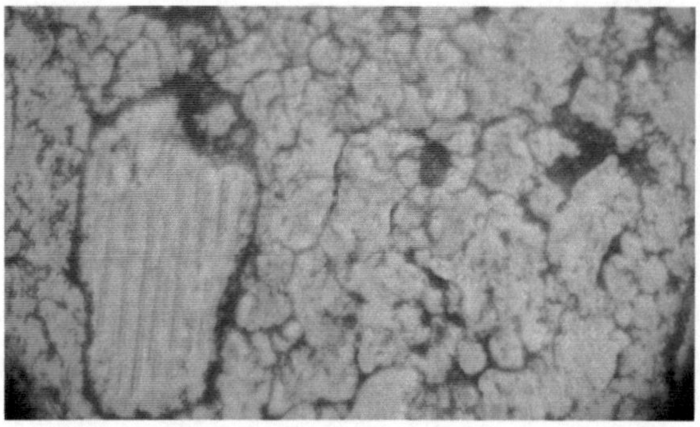

a

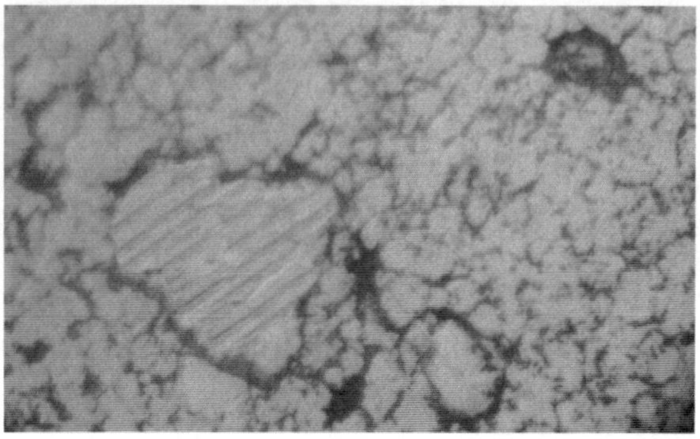

b

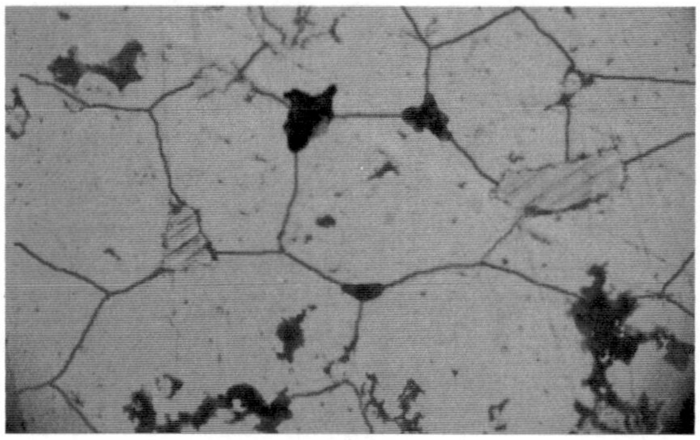

c

Figure 2. Microstructures of Fe-11.5wt%Cr sinter at a) 9000C b) 10000C c) 11000C, 25x

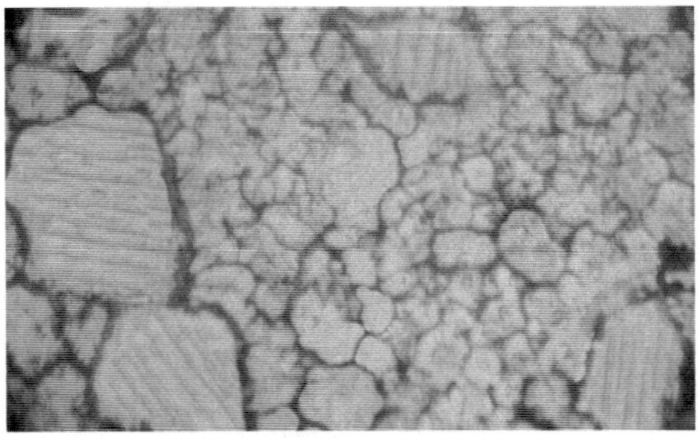

a

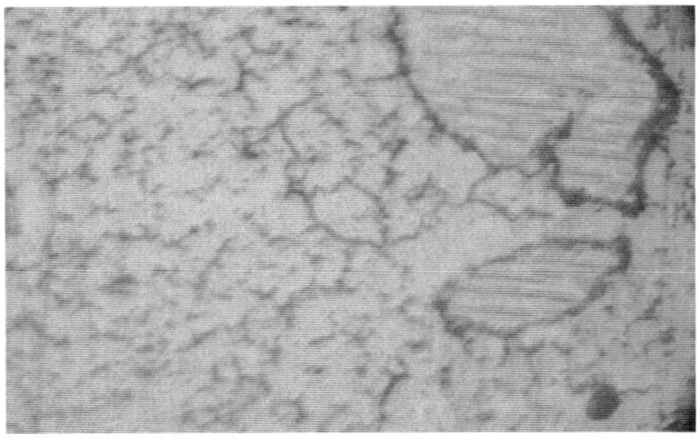

b

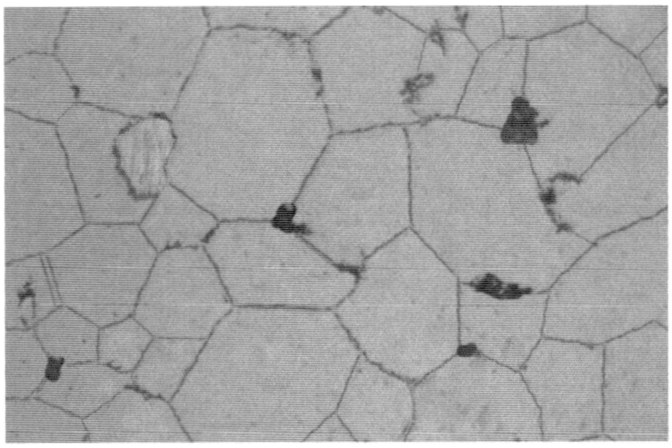

c

Figure 3. Microstructures of Fe-11.5wt%Cr-1wt%ZrO_2 sinter at a)900^{0}C b)1000^{0}C c)1100^{0}C, 25x

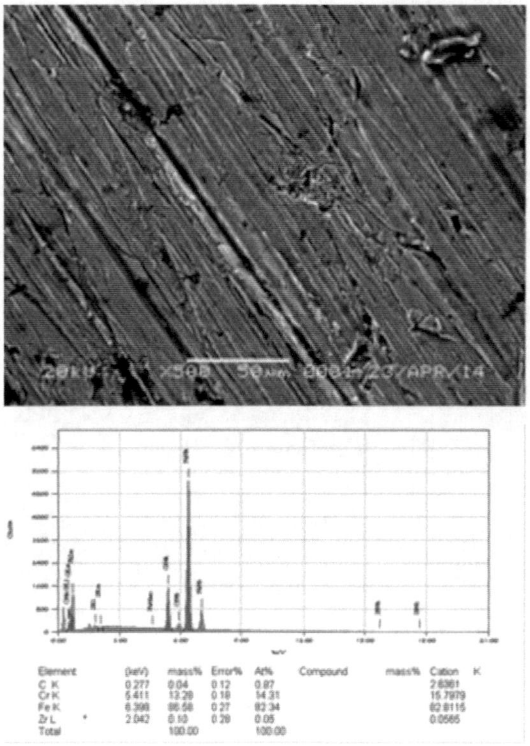

Figure 4. SEM-EDS Fe-11.5wt%Cr-1wt%ZrO2

Micro Vickers hardness measurements performed on Fe-11.5wt%Cr and Fe-11.5wt%Cr-1wt%ZrO_2 have shown in figure 5. It was found that the micro hardness value of Fe-11.5wt%Cr-1wt%ZrO_2 highest compare with micro hardness value of Fe-11.5wt%Cr. Average value of Hardness Vickers values was resulted 135.5 HV for Fe-11.5wt%Cr whereas 138.4 HV for Fe-11.5wt%Cr-1%wtZrO_2. It can be concluded that the addition nano zirconia increase micro hardness Fe-11.5wt%Cr-1%wtZrO_2 Sintering temperature is also increasing hardness of Fe-11.5wt%Cr and Fe-11.5wt%Cr-1wt%ZrO_2. Due to increasing density of the specimen.

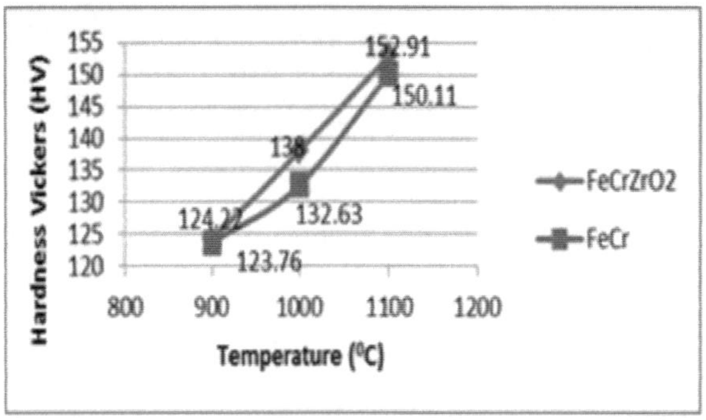

Figure 5. Micro Vickers hardness measurements

IOP Conf. Series: Journal of Physics: Conf. Series **799** (2017) 012018 doi:10.1088/1742-6596/799/1/012018

4. Conclusion

On the basis of the results the following conclusions can be as followed:

1. There are not significant differences in the microstructure between Fe-11.5wt%Cr and Fe-11.5wt%Cr-1wt%ZrO2. EDS examination showed that zirconia exit in the alloy Fe-11.5wt%Cr-1wt%ZrO2
2. An increase in the sintering temperature shift the microstructure from dendritic to equaxed.
3. The XRD pattern was found that ferritic and martensitic were the major phase and minor phase, respectively.
4. It was found that the highest micro hardness value was achieved by addition nano zirconia. Average value of Hardness Vickers values was resulted 135.5 HV for Fe-11.5wt%Cr whereas 138.4 HV for Fe-11.5wt%Cr-1%wtZrO2. wtZrO2 Sintering temperature is also increasing hardness of Fe-11.5wt%Cr and Fe-11.5wt%Cr-1wt%ZrO2. Due to increasing density of the specimen.

Acknowledgments
This research was supported by a grant from Jenderal Ahmad Yani University.

Reference

[1] 1. A. Méndez-Vilas and J. Díaz (Eds.), 2010 Microscopy: Science, Technology, Applications and Education, ©FORMATEX 2010 1811

[2] P. Fern´andez, A.M. Lancha, J. Lape˜na, R. Lindau, M. Rieth, M. Schirra, Creep strength of reduced activation ferritic/martensitic steel Eurofer'97 Fusion Engineering and Design 75–79 (2005) 1003

[3] E. Daum, K. Ehrlich, M. Schirra, Proccedings of the second milestone meeting on development of ferritic/martensitic steels for fusion technology, FZKA Report,FZKA, 1997, May, No. 5848.

[4] J. Lapen, M. Garcia-Mazarı´o, P. Ferna´ndez, A.M. Lancha ,Chemical segregation behaviour under thermal ageing of the low activation F–82H modified steel, J. Nucl.Mater. , 662, (2000), 283–287

[5] Benjamin J. S., Mechanical Alloying, *Scientific American, 234, (1976), 40-8.*

[6] Mukhopadhyay D. K., Froes F. H., Gelles D.S. Development of oxide dispersion strengthened ferritic steels for fusion, *Journal of Nuclear Materials*, Vol.1209, (1998) 258-263.

[7] Ohtsuka S., et al. , Nano-structure control in ODS martensitic steels by means of selecting titanium and oxygen contents, *Journal of Physics and Chemistry of Solids*, Vol. 66,(2005)571-575.

[8] Patil U., et al., An unusual phase transformation during mechanical alloying of an Fe-based bulk metallic, *Journal of Alloy and Compounds*, Vol. 389, (2005)121-126.

[9] Suryanarayana , Mechanical alloying and milling, *Progress in Mat. Science*, Vol. 46,(2001) 1-184.

IOP Conf. Series: Journal of Physics: Conf. Series **799** (2017) 012019 doi:10.1088/1742-6596/799/1/012019

Prediction of Melting Penetration of Armco Iron by Liquid Uranium Using MPS_LER

A P A Mustari[1*], Iksal[2], Sumiati[2], Syeilendra[1], Dwi Irwanto[1] and N A Aprianti[1]

[1]Nuclear Physics and Biophysics, Faculty of Mathematics and Natural Sciences, Institut Teknologi Bandung, Indonesia
[2]Faculty of Information Technology, University of Serang Raya, Jl. Raya Serang Cilegon Km 5, Serang, Banten, Indonesia.

[*]E-mail : pramutadi@fi.itb.ac.id

Abstract. Given the nuclear reactor severe accident, molten uranium may relocate and contact with iron based structural material which could result in melting of iron based material. According to binary phase diagram, contact between Fe and U may include eutectic phenomenon where melting temperature of the material drop sharply. Information of melting penetration rate of iron based material by uranium is important for melting relocation investigation during severe accident. MPS_LER has been developed and validated to simulate eutectic phenomena of Fe-U system. In the present study, the MPS_LER was used to predict melting penetration of Armco Iron by liquid uranium that has not been experimentally recorded. The present results show MPS_LER's prediction on melting penetration rate of Fe-U system at high temperature.

1. Introduction

Within the severe accident of Fukushima's reactors, good lessons were given in term of studying accident characteristics of reactor, melting relocation, design system, etc. Melting relocation is a complex phenomenon that is not simply a matter of melting point. During the accident, temperature increased rapidly while swelling of internal materials occurred that induced contact with other neighbouring components. Contact between particular materials could lead to formation of eutectic reaction that melt the materials far below their melting point. Thus, concern must be given to this mechanism to hinder underestimation of relocation process. Underestimation of the process occurs when calculation is based only on their melting point without taking into account eutectic reaction. On the other hand, investigation on the melting penetration of structural material, e.g. stainless steel, by molten fuel, i.e. uranium based material, is very rare [1]. The previous experiment regarding melting penetration of steel by molten uranium only provides results up to certain temperature. Relatively high temperature melting rate data is not yet provided.

To cope budget problem when conducting high temperature experiment for simulating severe accident, computer code of MPS method has been developed and equipped with eutectic reaction module namely MPS_LER. This module has been validated with TREAT experiment, an experiment addressed for severe accident investigation [2]. Additional validation with Pb-Sn system has also been successfully conducted [3].

In accordance, the objective of this study is to use equipped eutectic reaction module of MPS method in order to generate penetration rate of Armco iron by molten uranium at higher temperature than existing experimental data, i.e. 1300 °C.

2. Numerical methods

2.1. MPS method

The MPS method employs continuity, Navier-stokes and energy equations as governing equations,

$$\frac{D\rho}{Dt} = 0 \qquad \text{(equation 1)}$$

$$\rho\frac{D\vec{u}}{Dt} = -\nabla P + \mu\nabla^2\vec{u} + \rho\vec{g} \qquad \text{(equation 2)}$$

$$\frac{\partial H}{\partial t} = k\nabla^2 T + Q \qquad \text{(equation 3)}$$

Equation 1 shows the constant fluid density in order to implement the incompressible condition. The second equation, the Navier-stokes equation, employ pressure, gradient, viscous and gravity terms on the right hand side. Conduction and heat source terms are used in the energy equation (equation 3).

Weight function is employed to defined particle interaction with its neighbouring particles in the interaction radius,

$$w(r) = \begin{cases} \frac{r_e}{r_0} - 1 & (0 < r < r_e) \\ 0 & (r_e \le r) \end{cases} \qquad \text{(equation 4)}$$

where r is the distance between two particles and r_e is the effective radius. The particle number density of particle i is interpreted as

$$n_i = \sum_{j \neq i} w(|\mathbf{r}_j - \mathbf{r}_i|) = n^0 \qquad \text{(equation 5)}$$

where $\boldsymbol{r_i}$ and $\boldsymbol{r_j}$ are the position of particles.

The gradient and the Laplacian of differential operators are illustrated by the following particle interaction models,

$$\langle\nabla\Phi\rangle_i = \frac{d}{n^0}\sum_{j \neq i}\frac{\Phi_j - \Phi_i}{|\mathbf{r}_j - \mathbf{r}_i|^2}\left[(\mathbf{r}_j - \mathbf{r}_i)w(|\mathbf{r}_j - \mathbf{r}_i|)\right] \qquad \text{(equation 6)}$$

$$\langle\nabla^2\Phi\rangle_i = \frac{2d}{\lambda n^0}\sum_{j \neq i}\left[(\Phi_j - \Phi_i)w(|\mathbf{r}_j - \mathbf{r}_i|)\right] \qquad \text{(equation 7)}$$

Where d and n^0 are the number of spatial dimensions and initial particle number density, respectively.

$$\lambda = \frac{\sum_{j \neq i}|\mathbf{r}_j - \mathbf{r}_i|^2 w(|\mathbf{r}_j - \mathbf{r}_i|)}{\sum_{j \neq i} w(|\mathbf{r}_j - \mathbf{r}_i|)} \cong \frac{\int_V w(r)r^2 dv}{\int_V w(r)dv} \qquad \text{(equation 8)}$$

Differential operators in the governing equations are substituted by the particle interaction models, and then motion equations of the particles are obtained.

In the original MPS method, both the courant's stability and diffusion stability criteria are utilized to reach the convergent solution. The diffusion stability criterion, which is the numerical stability condition for viscous term, will restraint the maximum time step according to the value of kinematic viscosity. In other words, an increase of viscosity will result in decrease of maximum time step. The pressure calculation is done by solving the following Poisson equation of pressure,

$$\frac{1}{\rho_0}\nabla^2 P = -\frac{1}{\Delta t^2}\frac{n^* - n^0}{n^0} \qquad \text{(equation 9)}$$

where $\boldsymbol{\rho_0}$ and $\boldsymbol{n}^*$ are the temporal density and initial particle number density, respectively.

2.2. *Eutectic reaction calculation model*

In this model, the definition of eutectic melting is the condition where melting occur below its melting point due to mixing of the two materials. The modelling of eutectic phenomenon is consisted of two steps. The first step is mass transfer of ions among particles. The solution is provided by solving Fick's second law of mass diffusion. The second step is the application of boundary based on binary phase diagram as chemical reaction.

The first step is the utilization of Fick's second law as governing equation for mass diffusion by

$$\frac{\partial m}{\partial t} = \mathrm{D}\nabla^2 m \qquad \text{(equation 1)}$$

where m is mass and D is diffusion coefficient of material. Mass diffusion term of right side of the equation (equation **1**) is calculated by MPS Laplacian model explicitly.

The second step is correlating mass diffusion result in a particle with eutectic criteria from binary phase diagram. In the phase diagram, the criteria are given in temperature and mass fraction. Detail explanation of the model, MPS_LER, is explained elsewhere [3].

2.3. *Simulation condition*

In this simulation, the minimum particle distance was about 0.0001 m. The test section thickness of the MPS geometry was about 500 μm. The total number of particles for developing the test geometry was 42164. Figure 2 shows the 2D calculation model for TREAT experiment. The orange color area is the molten uranium and blue color is the tested material, i.e. Armco iron. Grey color act as sauresin and the wall tube, which is wall particle in the MPS method simulation.

Diffusion coefficient was obtain from first principle molecular dynamics (FPMD) code VASP simulation [4]. The calculation was conducted on metal fuel U-Pu-Zr and cladding Fe. Detail information of diffusion coefficient calculation of SS304 and uranium is explained in previous work [2].

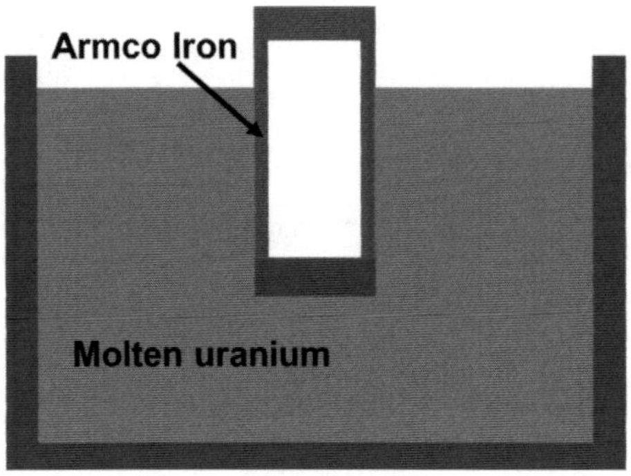

Figure 2 MPS geometry adapted from TREAT experiment..

3. Results and discussion

Figure 2 shows melting penetration pattern of molten uranium into test section of Armco iron. The result shows the eutectic reaction where the steel transform into liquid phase. At 1360°C, Armco iron starts to melt at 0.48 s of about 0.0001 m (particle diameter) as shown in figure 2. At 1390°C, the steel test section melts faster than previous temperature at 0.36 s with the same thickness. Outer layer melted first as ion diffusion proceed between uranium particles and outer Armco iron particles. Motion of molten uranium near the test section was due to temperature gradient along vertical axis.

Figure 3 shows data of melting penetration rate vs. temperature. Blue circle indicates the result from TREAT experiment, while red circle indicates previous result of MPS method for validating the eutectic calculation model [2]. The green circle demonstrate calculation result of MPS_LER method for higher temperature than recorded experimental data. It is found that the penetration rate value at 1360°C was slightly higher than MPS result at 1300°C. And result at 1390°C shows bigger number for melting penetration rate. The phenomena occurs due to increase of diffusion coefficient with increase of temperature. Another reason is that required mixing ratio (uranium content) become smaller for temperature near the Armco iron's melting point.

The melting penetration data that have not been recorded nor conducted in the experiment will give valuable information for further study of melting relocation during severe accident.

Figure 1 melting penetration pattern at 1360 and 1390 °C.

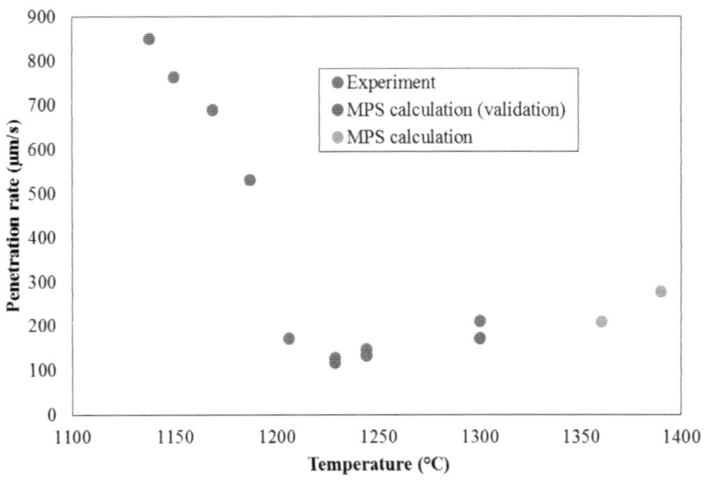

Figure 2 Results of MPS calculation prediction at high temperature including data of validation of MPS_LER.

4. Conclusion

MPS_LER was succesfully calculate melting penetration at temperature higher than 1300°C. Two temperature conditions were calculated, i.e. 1360 and 1390 °C. Melting penetration rate of both temperatures showed enhancement. The condition was strongly influenced by increase of diffusion coefficient and decrease of the required diffuse iron content at high temperature.

Acknowledgments

Highly appreciation to Prof. Koshizuka and Prof. Shibata from Tokyo University for providing MPS code.

References

[1] Walter C M and Dickerman C E 1964 Treat Study of the Penetration of Molten Uranium and U / 5 wt % Fs Alloy through Type 304 Stainless Steel * *Nuclear science and engineering* **18** 518–24

[2] Mustari A P A and Oka Y 2014 Molten uranium eutectic interaction on iron-alloy by MPS method *Nuclear Engineering and Design* **278** 387–94

[3] Mustari A P A, Oka Y, Furuya M, Takeo W and Chen R 2015 3D simulation of eutectic interaction of Pb-Sn system using Moving Particle Semi-implicit (MPS) method *Annals of Nuclear Energy* **81** 26–33

[4] Himi M, Yamamoto Y, Nagamine Y, Shirakawa N, Uehara Y and Arima T 2009 Next generation safety analysis methods for SFRs (8) analysis of eutectics between fuel and steel in metal fuel with FPMD code VASP *ICONE17-75578* (Brussels, Belgium: ASME) pp 1–10

Meso-Scale Radioactive Dispersion Modelling using GPU

Sunarko[1*], **Zaki Suud**[2]

[1]BATAN, Jl. Kuningan Barat, Jakarta 10270, Indonesia
[2]ITB, Jl. Ganesha 10, Bandung 50254, Indonesia

*Email: sunarko@batan.go.id

Abstract. Lagrangian Particle Dispersion Method (LPDM) is applied to model atmospheric dispersion of radioactive material in a meso-scale of a few tens of kilometers for site study purpose. Empirical relationships are used to determine the dispersion coefficient for various atmospheric stabilities. Diagnostic 3-D wind field is created based on data from a meteorological station using mass-conservation principle. Particles imitating radioactive pollutant are dispersed in the wind-field as a point source. Time-integrated air concentration is calculated using kernel density estimator (KDE) in the lowest layer of the atmosphere. Parallel code is developed for GTX-660Ti GPU with a total of 1344 scalar processors using CUDA programming. Significant speedup of about 20 times is achieved compared to the serial version of the code while accuracy is kept at reasonable level. Only small differences in particle positions and grid doses are observed when using the same sets of random number and meteorological data in both CPU and GPU versions of the code.

1. Introduction

Lagrangian Particle Dispersion Method (LPDM) is now commonly used in atmospheric dispersion studies. This method uses hypothetical particles to represent substances undergo advection by wind and diffusion process in the atmosphere as well as deposition to the ground. At any time, these particles will reside in a certain layer of the atmosphere and concentration calculation should be made for father analysis. Kernel Density Estimator (KDE) is one of many methods that can be employed to provide estimation for the concentration. In application where time-integrated concentration is required such as for dose calculation in the dispersion of radioactive substances, running time becomes crucial because concentration calculation should be performed over a specified interval.

The use of GPU arised from the demand for real-time, high-definition 3D graphics has driven the development of programmable Graphic Processor Unit or GPU into a highly parallel, multithread, many-core processor with tremendous computational power and high memory bandwidth. In the past decade, parallel computation using Graphical Processing Unit (GPU) has gained broader acceptance in scientific computation including dispersion modeling. The use of GPU has grown exponentially due to the appearance of new architectures and programming models such as NVIDIA CUDA (formerly abbreviated from Compute Unified Device Architecture[7]. Compared to a multicore CPU, a GPU can contain much larger number of processors and can provide speedups of more than 100 times in science and engineering applications[3]. This advantage can be utilized in KDE calculation since it is inherently parallel. Computation in a grid point can be done independent of other grid points each with the same set of particles.

In this paper we present a parallel code for radioactive dose calculation from atmospheric dispersion of radionuclides using a GPU and measure its performance improvement over a single-threaded code for CPU application.

2. Model description

There are several main processes in the model necessary for facilitating particle dispersion such as wind-field preparation, elevation model, dispersion coefficient, concentration calculation method and random number generation. The flow of process in the code is provided in Figure 1.

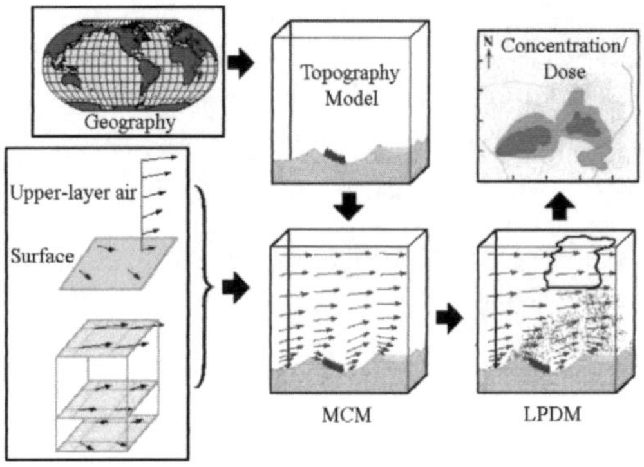

Figure 1. Process flow[1]

Lagrangian Particle Dispersion Method (LPDM) is run by tracking a number of particles in a flow field. LPDM uses stochastic differential equation to explain the similar process as the advection-diffusion equation in Lagrangian framework**Error! Reference source not found.**. Stochastic differential equation for the movement of ideal fluid in three dimensions is:

$$dx = \bar{u}dt + (2K_x)^{\frac{1}{2}}dW \quad (1)$$

$$dy = \bar{v}dt + (2K_y)^{\frac{1}{2}}dW \quad (2)$$

$$dz = \bar{w}dt + \frac{\partial K_z}{\partial z}dt + (2K_z)^{\frac{1}{2}}dW \quad (3)$$

where $dW_{x,y,z}$ is a random number with zero average and variance dt, namely $d\overline{W} = 0$ and $d\overline{W}^2 = dt$. In equation (4) and (5) it is assumed that turbulence is homogeneous in x and y directions. The above equation can be integrated with time to obtain particle path which represent the movement of each individual particle. In numerical calculation of dispersion, the number of released particle is large, each with its own label and certain mass or activity. Concentration at time t can be calculated from locations of particles at time t and particle activity.

Implementation of random-walk method is explained for vertical direction in**Error! Reference source not found.**. The equation for horizontal direction has the same form with the vertical one but without the differential form of eddy diffusivity (K). The vertical and horizontal dispersion coefficient is taken from Caughey et al. [11] for various stability conditions.

Three-dimensional meteorological data is not always available from meteorological agencies. Frequently, site-specific meteorological data should be obtained by way of one or more monitoring stations with limited height and parameters. Data is usually recorded over a certain period of time,

[1] Figure adapted from: https://str.llnl.gov/str/Baskett.html

often hourly. Consequently, the 3-dimensional wind-field should be prepared hourly for the whole duration of the simulation. Three dimensional wind field is prepared using mass-conservation principle such as the Mass Consistent Model (MCM) [2] which is sufficient for modeling up to the meso-scale level. This method is capable of utilizing data from a limited data from a meteorological tower while at the same time provide usable data for particle advection. The wind field is produced in the CPU and then transferred to the 3D texture memory in the device before being used by the kernel. MCM was proposed by Sasaki in 1958 and implemented by Sherman in 1978. MCM employs minimization of functional:

$$E(u,v,w,\lambda) = \int_V [\alpha_1^2(u-u^0)^2 + \alpha_1^2(v-v^0)^2 + \alpha_2^2(w-w^0)^2 + \lambda(\frac{\partial u}{\partial x} + \frac{\partial v}{\partial y} + \frac{\partial w}{\partial z})]dx\,dy\,dz \quad (4)$$

Equation (4) is solved using finite-difference method in a 101×101×31 rectangular grid to obtain 3-dimensional wind field. The spacings of the grid are $\Delta x = 1\ km$, $\Delta y = 1km$, $\Delta z = 50$m. Air is assumed as incompressible and meteorological data from one station is used (u^0, v^0, w^0). At convergence, the wind field at grid points (u, v, w) are obtained. The values for α_1 and α_2 are determined by the atmospheric stability category. The lagrange multiplier λ is used to define the boundary condition.

After a number of particles undergo transport and diffusion process and their locations are noted, the concentration at a certain location and time can be determined by counting the number of particles on a sampling area centered at the grid point (x,y,z) known as box-counting method as follows:

$$C(x,y) = \frac{1}{\Delta x\ \Delta y \Delta z}\sum_{i=1}^{N} m_{pi} I \quad (5)$$

$$I = \begin{cases} 1 \quad if \quad \begin{matrix} |X_i - x| < \Delta x/2 \\ |Y_i - y| < \Delta y/2 \\ |Z_i - z| < \Delta z/2 \end{matrix} \\ 0 \qquad others \end{cases} \quad (6)$$

m_{pi}, X_i, Y_i, Z_i are mass or activity in the coordinate (X, Y, Z) of particle i. Concentration is calculated in an area $\Delta x \Delta y \Delta z$ with N number of particles. The increase in N will increase computation time and an increase in sampling area will decrease the resolution.

Computation efficiency can be increased using KDE **Error! Reference source not found.**. Concentration on a grid point can be calculated from the sum of contribution from each particle:

$$C(x,y) = \sum_{i=1}^{N} \frac{m_{pi}}{h_{xi}h_{yi}h_{zi}} K(r_{xi}, r_{yi}, r_{zi}) \quad (7)$$

$$(1/h_{xi}h_{yi}h_{zi}) \int_{-\infty}^{\infty}\int_{-\infty}^{\infty} K dx dy dz = 1 \quad (8)$$

$r_{xi} = (X_i - x)/h_{xi}$, $r_{yi} = (Y_i - y)/h_{yi}$, $r_{zi} = (Z_i - z)/h_{zi}$ are distances of particle (X_i, Y_i, Z_i) with grid (x, y, z). Parameters h_x, h_y, and h_z are bandwidths to determine the degree of smoothness on each coordinate. Average concentration can be calculated using a simple uniform kernel as follows:

$$K(r_x, r_y, r_z) = \frac{1}{8} I_x I_y I_z \quad (9)$$

$$I_{x,y,z} = \begin{cases} 1 & r_{x,y,z}^2 < 1 \\ 0 & others \end{cases} \quad (10)$$

Bandwidth is proportional to the grid space used: $h_x = \alpha\Delta x$, $h_y = \alpha\Delta y$, $h_z = \alpha\Delta z$. $\alpha = 1.0$ is used to obtain smooth concentration profile. A particle can contribute to the concentration on several grid points at a time.

3. Methodology

MODAL will be used to calculate external gamma exposure from I-131, Cs-137, Kr-88 and Xe-133 isotopes emitted through a 50 meter high stack in 6-hour duration. The modeled area is 100 km × 100 km with Cartesian grid size of $\Delta x = 1$ km, $\Delta y = 1$ km and $\Delta z = 0.05$ km. Total height of the model is 1500 meters. The air is incompressible and has uniform density anywhere in the grid.

Simulations are performed using isotopes of Cs-137, I-131, Kr-88, and Xe-133 each with a total strength of 0.011, 15.04, 62.000, and 8.37 million Curies respectively**Error! Reference source not found.**. The plume is released from a 50 m high stack at a constant rate from second 8.500 to 30.000 after the actual accident happens. Two scenarios are prepared each containing different amount of Xe-133 release provided in **Table 1**. For each isotope, maximum value is taken from available data for conservative assumption.

A number of simulations are performed representing dry and rain condition with 5, 10, 15 and 20 mm/hr rain intensities. Worst-case meteorological condition is picked from meteorological database from on-site monitoring program performed during site feasibility study in 1994-1995 periods. Worst-case condition in this paper is defined as condition where wind is blowing to a consistent sector during the entire duration of radioactive release. This type of atmospheric condition will tend to produce higher TIAC for locations in the downwind direction from the source emission. Frequency analysis is performed to pick a period in the database where wind blow consistency is found.

Parallel application

The CUDA programming model allows the launching of hundreds to thousands of GPU threads. CUDA comprises a programming language accompanied by the necessary libraries to support the execution of code on NVIDIA GPUs. A program that uses CUDA must be structured on kernels. These kernels are destined to run on the GPU (device) while the main routine is running on the CPU (host). In order to initiate a kernel execution using CUDA, one must define a mapping of application threads into blocks and accordingly a mapping of such blocks on a grid. Device code is executed by many independent threads, which are hierarchically organized; groups of threads are organized into grid. Within a block, 32 threads are organized into a group referred to as warp. Each streaming multiprocessor (MP) schedules and executes threads at the warp level[8]. A warp is the minimum scheduling unit. The device executes an instruction for all threads within a warp before launching the next instruction. Threads within a warp physically execute the same instruction in each cycle. The best performance is achieved when the number of threads in a thread-block is a multiple of the warp size[7].

CPU and GPU have their own separated memory and therefore data that is used by the GPU should be transferred first to the GPU before calculation can take place. The result of the calculation at the end of the process should be copied back to the CPU after the process in GPU is finished.

All compute devices follow the IEEE 754-2008 standard for binary floating-point arithmetic with a number of deviations. The existing discrepancy in floating-point capability between CPU and GPU is because GPU is specialized for compute-intensive, highly parallel computation which devotes more transistors to data processing rather than data caching and flow control[4]. It is important that this discrepancy does not cause large deviation in the calculation result between CPU and GPU systems.

Implementation

Wind-field, elevation model, dispersion coefficient and random number generations are performed by the CPU using mass-consistent method, Shepherd's method, empirical relationships and Marsenne-Twister algorithm respectively. Particle dispersion, concentration calculation and integration processes are performed by the GPU due to the large number of particles and grid coordinates being used.

The particles are released in plume segments or puffs with duration of a number of seconds and therefore one hour of simulation may be comprised of several segments. One segment is propagated based on the particle locations of previous segment to save computation time. This way the texture cache memory can be fully utilized because particle positions in a puff are more localized in space. Each particle carries position, lifetime and status information. The activity is proportional to the total release of the hour divided by the number of particle released during the hour and undergoing radioactive decay as the time progresses. At the end of each plume travel, concentrations are taken and time-integrated concentration is obtained for each grid point. Concentrations are calculated using kernel density estimator for the lowest layer of the atmosphere where effect to human health is the most significant.

The dispersion coefficients (K_x, K_y) are calculated in the GPU to avoid errors from interpolation when 1D texture memory is used to store CPU generated tables.

Random numbers used in the dispersion part of the simulation is generated in the CPU for every plume step and will be copied to the GPU prior to usage. The routine is performed using Marsenne-Twister library in C[10] with a period of 2^{19937}.

Parallel processing is performed for particle dispersion and concentration calculation. In particle dispersion process, advection vector and dispersion coefficients are read from the 3D and 1D textures respectively using current particle position. Each stream processor is responsible for one particle at a time and there are no interactions among particles. Particle positions, random numbers, dry and wet deposition coefficients are first copied in coalescent from general memory to shared memory in BLOCKSIZE chunks for faster processing. The process is repeated as necessary until all particles are processed for the current puff step. New particle positions are stored back to the general memory for concentration calculation.

In the concentration calculation, each stream processor is assigned to a grid coordinate and calculation is done using the position of particles produced in the earlier process. Due to the limited capacity of shared memory, particles are copied to shared memory and used in the calculation in stages each of BLOCKSIZE chunks. These particles are shared by the grid coordinates in the same block in the calculation.

4. Results and discussion

Comparison of calculation result between CPU and GPU is performed by comparing the distance between particles. In this case, a same set of random numbers is used in both version of the code. For every hour, 18000 particles are released in groups of puff. The duration of each puff is 60 seconds and therefore there are 60 puffs in one hour, each containing 300 particles which are further organized into 6 smaller groups containing 50 particles. The first group released will travel the furthest and will likely are further separated than subsequent groups. The distance between particles in each group is calculated and averaged and their values are provided in Figure 4. Simulated wind condition containing fluctuation of wind data is used. Minor numerical error arises from the differences in mathematical functions differences between CPU and GPU[1]. However, this small error may translate into larger numerical error because of the difference in wind vectors being used in the subsequent steps. The more complex the wind pattern, the larger is the expected deviation.

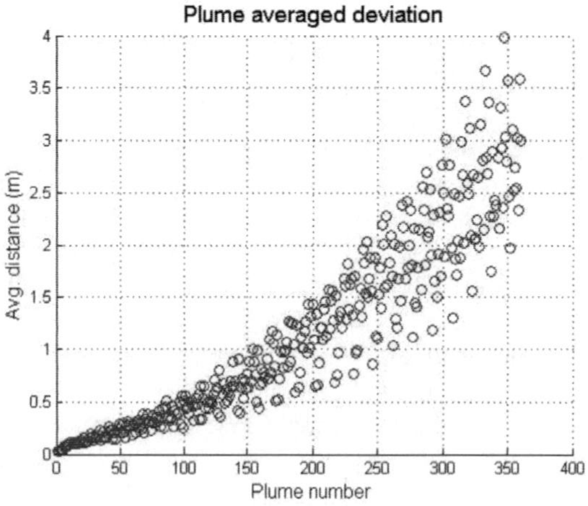

Figure 2. Particle separation for each particle group.

The effect of particle separation will be carried forward to the concentration at grid points. Concentrations are calculated using Kernel Density Estimation for the lowest 5 meter layer because it is the most important layer in dispersion calculation. Calculations are performed for 101 x 101 grid points and the result of non-zero grid dose is provided in Figure 4. The above and below lines are plus and minus one standard deviation respectively.

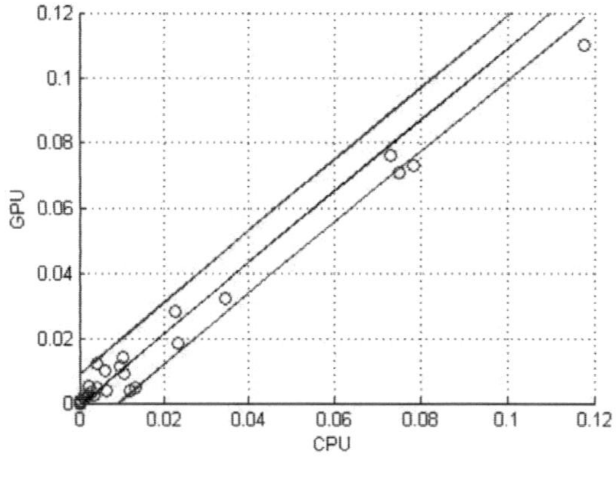

Figure 3. Scatter plot of doses at grid points.

Using a linear fit, it is found that the dose resulting for GPU calculation is about 15.6% higher. This is because more particles are dispersed from emission height into the lowest layer of the atmosphere where concentration calculations are performed. Dispersion and TIAC are calculated by the GPU involving trilinear interpolation of wind field, dispersion coefficient and concentration calculations. Only random number generation and wind field calculation are performed by the CPU.

Performance.

Simulations with different number of particles are also performed on a 3.0 GHz Core i5-3330 with 2 GB of RAM and GPU GTX660Ti with 1344 scalar processor and 2 GB memory. Figure 5 shows the speedup gain using the GPU. Speedup is contributed by many factors such as memory bandwidth, faster access to constant and shared memory as well as memory caching in GPU. Acceleration is also provided by the readily available trilinear interpolation function in CUDA when 3D texture memory is used[1].

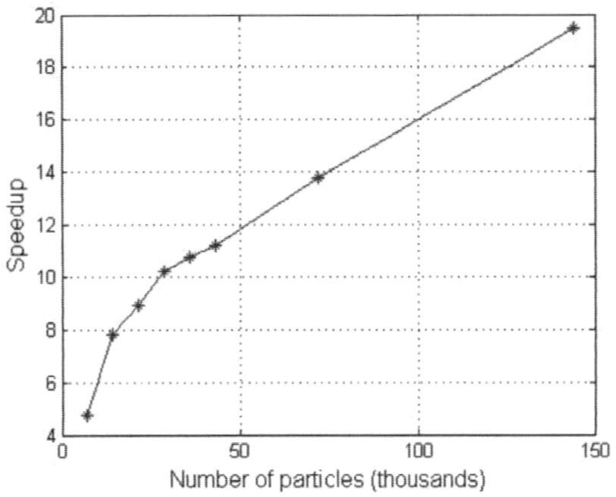

Figure 4. Speedup by GPU application.

The speedup achieved is increasing with the increasing number of particles in use. In this experiment, approximately linear growth in speedup can be identified after 28 800 particles or more are used. CPU calculation was stopped at 144 000 particles/hour because of limited memory capacity.

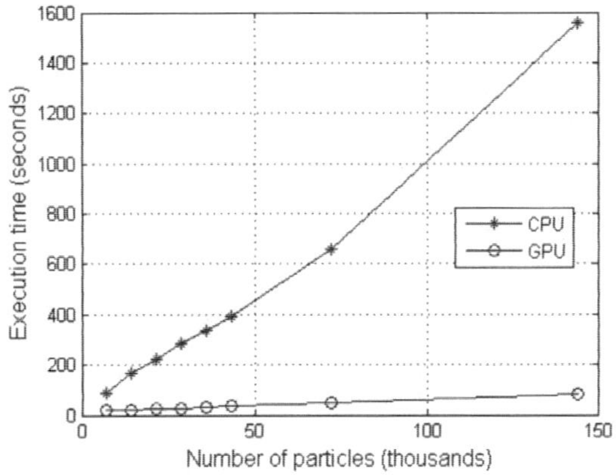

Figure 5. Running time.

Molnar measure performance without performing concentration integration. In this paper, concentration integration is performed every 60 second to calculate the time-integrated air concentration (TIAC) at each grid point which directly represents the dose received. Due to the limitation in shared memory capacity, the random number used in dispersion calculation is being read directly from general memory which slows down running time.

Zeng et al. [9] has also performed a parallelization based on another Lagrangian particle dispersion method named FLEXPART using 2.0 GHz and 4 GB CPU and comparing the performance against Tesla C1060 1.3 GHz and 4 GB memory GPU and achieved an acceleration of about 10 times.

5. Conclusion

GPU has been found to be a cost-effective solution in accelerating computational power for LPDM. Parallel version LPDM has achieved multiplication factor of about 20 and will increase with the increasing number of particles used. Differences in computation method between CPU and GPU systems as indicated by separation of particles at the end of 1 hour simulation may reach as far as 4 meters when the same set of random numbers and wind-field are used. The doses at grid points are slightly higher for computation using GPU by about 15.6% as a result of more particle dispersion into concentration calculation layer.

Acknowledgment

The authors acknowledge the Ministry of Research and Technology and Nuclear Physics Laboratory of Institute Teknologi Bandung for supporting this research.

References

[1] Molnar Jr. F, Szakaly T, Meszaros R and Lagzi L 2010 *Computer Physics Communication* **181** 105.
[2] Sherman C A 1977 *Journal of Applied Meteorology* **17** 312.
[3] NVIDIA Corporation 2010 *NVIDIA CUDA programming Guide Version 2.3.1*.
[4] NVIDIA Corporation 2010 *NVIDIA CUDA C programming Guide Version 4.2*.
[5] Uliasz M 1994 *Lagrangian Particle Dispersion Modeling in Mesoscale Applications. Environmental Modelling* II. ed. P. Zannetti (Southampton, Boston: Computational Mechanics Publication p 71)
[6] Sunarko 2009 *Model dispersi partikel untuk perhitungan lepasan radioaktif (Studi kasus semenanjung Muria-Jawa Tengah)* (Institut Teknologi Bandung : Master-Thesis).
[7] Jimenez J and Ruiz de Miras J 2012 *Computer Methods and Programs in Biomedicine* **108** 1229.
[8] Kalantzis G and Tachibana H 2014 *Computer Methods and Programs in Biomedicine* **113** 116.
[9] Zeng J, Matsunaga T and Mukai H 2010 *Using NVIDIA GPU for Modelling the Lagrangian Particle Dispersion in the Atmosphere, International Environmental Modelling and Software Society (iEMSs)* (Ottawa, Canada: International Congress on Environmental Modelling and Software Modelling for Environment's Sake, Fifth Biennial Meeting).
[10] Matsumoto M and Nishimura T http://www.math.sci.hiroshima-u.ac.jp/~m-mat/MT/emt.html.
[11] Caughey S J, Wyngaard J C and Kaimal J C 1979 *J. Atmos. Sci.* **36** 1041.

IOP Conf. Series: Journal of Physics: Conf. Series **799** (2017) 012021 doi:10.1088/1742-6596/799/1/012021

Study Neutronic of Small Pb-Bi Cooled Non-Refuelling Nuclear Power Plant Reactor (SPINNOR) with Hexagonal Geometry Calculation

Dwita Nur Krisna[*], **Zaki Su'ud**
Nuclear Physics and Biophysics Research Division, Physics Department, Faculty of Mathematics and Natural Science, Bandung Institute of Technology Indonesia
Jl. Ganesha 10, Bandung 40132, Indonesia

E-mail: dwitakrisna@gmail.com

Abstract. Nuclear reactor technology is growing rapidly, especially in developing Nuclear Power Plant (NPP). The utilization of nuclear energy in power generation systems has been progressing phase of the first generation to the fourth generation. This final project paper discusses the analysis neutronic one-cooled fast reactor type Pb-Bi, which is capable of operating up to 20 years without refueling. This reactor uses Thorium Uranium Nitride as fuel and operating on power range 100-500MWtNPPs. The method of calculation used a computer simulation program utilizing the SRAC. SPINNOR reactor is designed with the geometry of hexagonal shaped terrace that radially divided into three regions, namely the outermost regions with highest percentage of fuel, the middle regions with medium percentage of fuel, and most in the area with the lowest percentage. SPINNOR fast reactor operated for 20 years with variations in the percentage of Uranium-233 by 7%, 7.75%, and 8.5%. The neutronic calculation and analysis show that the design can be optimized in a fast reactor for thermal power output SPINNOR 300MWt with a fuel fraction 60% and variations of Uranium-233 enrichment of 7%-8.5%.

1. Introduction

As the times, many nuclear technologies are increasingly prevalent developed into renewable technology, particularly in the development of the Nuclear Power Plant (NPP) [1]. Nuclear power plant should be able to overcome the problems encountered so far, such as the urge to improve nuclear safety, radioactive waste issues, the cost of investment and operation of reactors which are expensive, and the danger of proliferation of nuclear weapons.

During 1990 and 2000 a lot of innovation nuclear reactors have been developed, one of the most prospective is the kind of fast reactors cooled Lead-Bismuth (Pb-Bi). Over time, this time in various countries especially Japan has developed rapidly cooled reactor type Lead and Lead-Bismuth mixture. It is done to seek the weakness of the use of cooling agents such as sodium that are more explosive when interacting with water and air, such as the Japanese Monju reactor accident in 1999 [2,3].

SPINNOR reactor is cooled nuclear reactor concept of Lead-Bismuth with fast neutron spectrum that can be operated for long years without refueling. SPINNOR reactor is considered as a concept that has a high priority in the development of nuclear energy systems. SPINNOR concept focus on sustainable nuclear energy, efficient use of resources, minimal waste, and inherent safety [1].

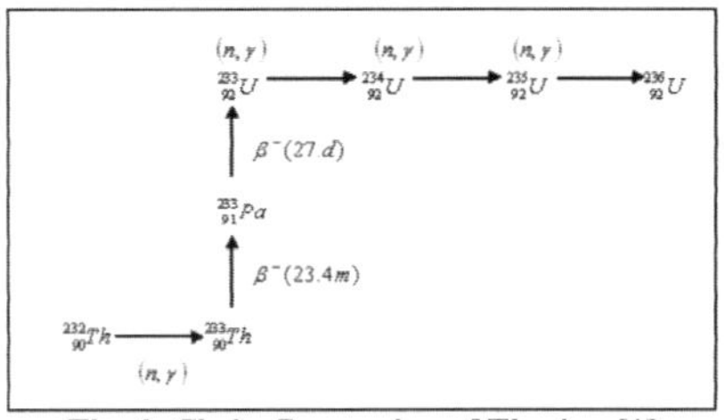

Fig. 1. Chain Conversion of Thorium[1]

Thorium produces 90% less waste compared to Uranium, and requires only about 200 a year to store the waste than uranium which requires 10,000 years. In the universe, we can say all are Th-232 which has a

[*] Corresponding author.
E-mail address: dwitakrisna@gmail.com

half life of about 14.05 billion years[1]. In this study, it will be designed reactors SPINNOR or type fast reactor cooled Pb-Bi, will now be analyzed neutronic reactor SPINNOR fueled Thorium Uranium Nitride in 20 years of operating time, and will be calculated the optimum design of SPINNOR reactor.

2. Design Concept and Calculation Methods

From the data presented in table 1, it is obtained several plots a graph of calculation results Pij burn-up during reactor operation

Table 1. Variance in reactor design parameter survey of SPINNOR

Percentage of fuel (%)	*Fuel, Cladding,* and *Coolant* Fraction (%)	Power thermal (MWt)
Case A: Region F1/F2/F3 7% / 7.75% / 8.5%	50% *fuel*, 20% *cladding*, 30% *coolant fraction*	100
		200
		300
		400
		500
	55% *fuel*, 20% *cladding*, 25% *coolant fraction*	100
		200
		300
		400
		500
	60% *fuel*, 20% *cladding*, 20% *coolant fraction*	100
		200
		300
		400
		500

Fuel used in this study based on Thorium enrichment that comes from U-233. Thorium, that has the ability to produce fissile material, is significant in this U-233 case.

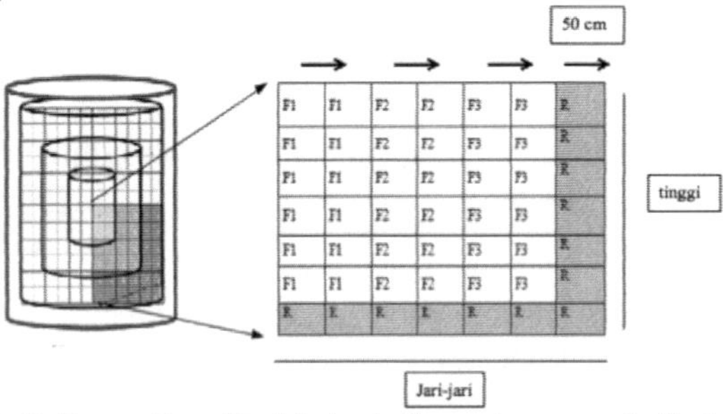

Fig. 2. Composition of fuel design in the reactor core and reflector

The calculation is performed by first making the system design modeling work SPINNOR reactor by using a computer simulation program SRAC (Standard Thermal Reactor Analysis Code) and using Ms. Excel for calculation of atomic density and power level. Calculation phase begins with the creation of the input file to the program SRAC for the calculation of burn-up with a variation of the percentage of uranium-233: Region Fuel1 / Fuel2 / Fuel3 by: 7%, 8.75%, 8.5%. The calculation using SRAC can be made in the form of a flowchart as follows:

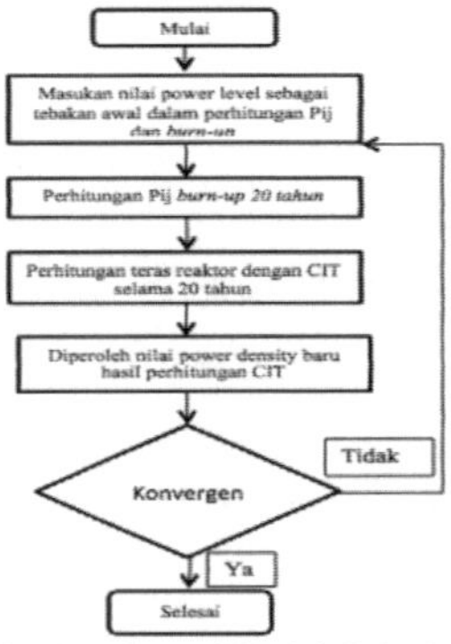

Fig. 3. Flowchart of SRAC Calculation

3. Results and Discussion

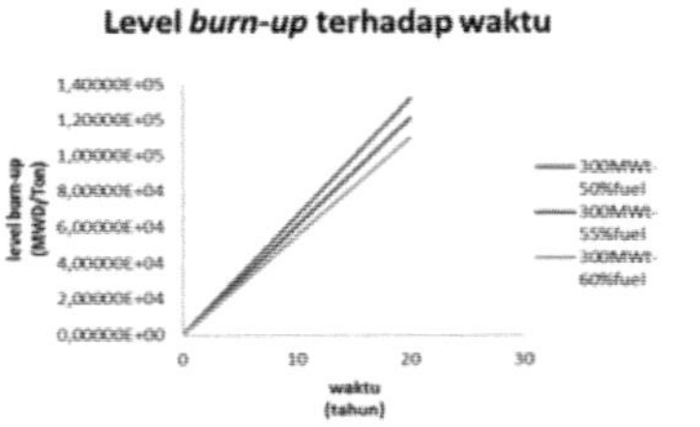

Fig. 4. Curve burn-up design with percentage of U-233 7%, 300MWt thermal power output, and variance of fuel fraction 50%, 55%, and 60%

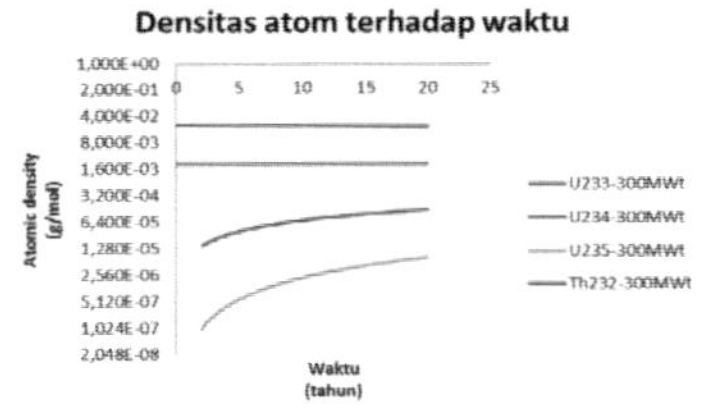

Fig. 5. Curve of atomic density changes over a period of burn-up on SPINNOR reactor with 300MWt thermal power output

Burn-up Calculation

Figure 4 show the curves burn-up level of the burning time for 20 years on output power variation 300MWt the fuel fraction 50%, 55%, 60%, and variations in the percentage of U-233 by 7%. The graph can be seen that the smaller fraction of fuel used, the greater value of the level of burn-up. It shows that the smaller fraction of fuel the less amount of fuel for the reactor core size equal to the third variation of the fuel fraction. Therefore, the value of the level of burn-up fraction is inversely proportional to the increase in fuel, so that the value of the level of high burn-up explaining that the fuel to be used effectively.

Atomic Density Analysis

The process of fuel burn-up greatly affects the density of the atoms making up the fuel. The density of atoms (N) is the concentration of atoms per unit volume. Figure 5 ilustrate the changes in the density of the atoms during the burn-up on SPINNOR reactor with a thermal output power variation at 300MWt and fuel fraction 60%. On the graph shows that over time the operation of the reactor, the number density of atoms of each nuclide will decrease.

Calculation of Effective Multiplication Factor and Reacivity

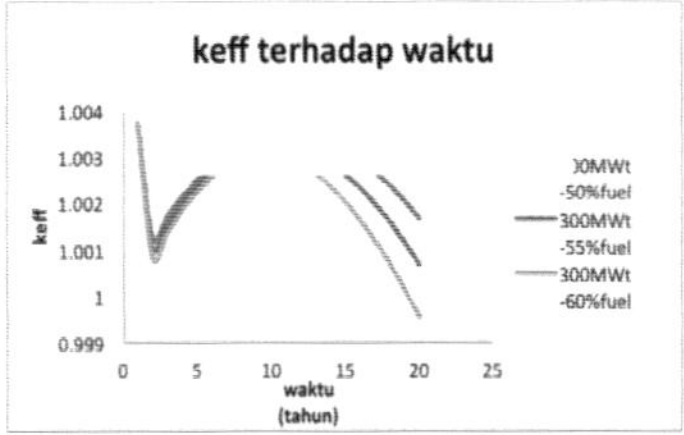

Fig. 6. Curve of k_{eff} in SPINNOR reactor with thermal power output 300MWt and variance fuel fraction 50%, 55% and 60%

time (year)

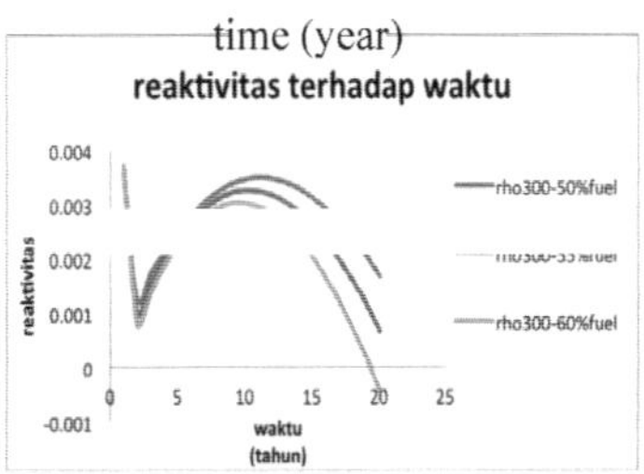

Fig. 7. Curve of reactivity on SPINNOR reactor with thermal power output 300MWt and variance fuel fraction 50%, 55%, and 60%

time (year)

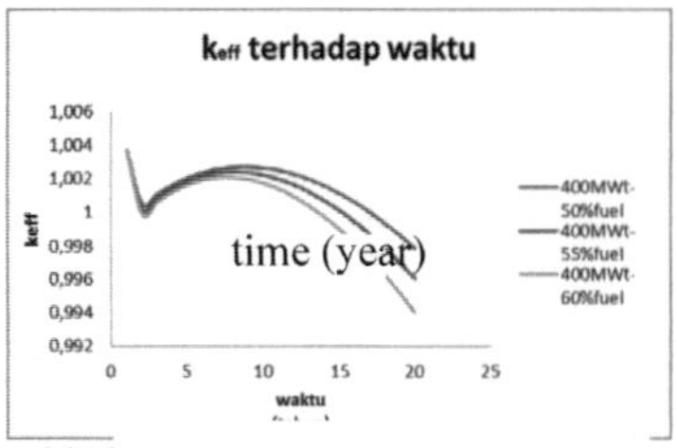

Fig. 8. Curve of k_{eff} in SPINNOR reactor with t[illegible] variance fuel fraction 50%, 55% and 60%

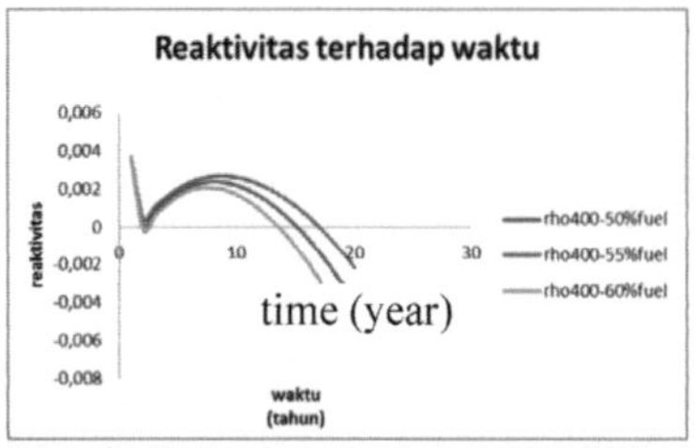

Fig. 9. Curve of reactivity on SPINNOR reactor with thermal power output 400MWt and variance fuel fraction 50%, 55%, and 60%

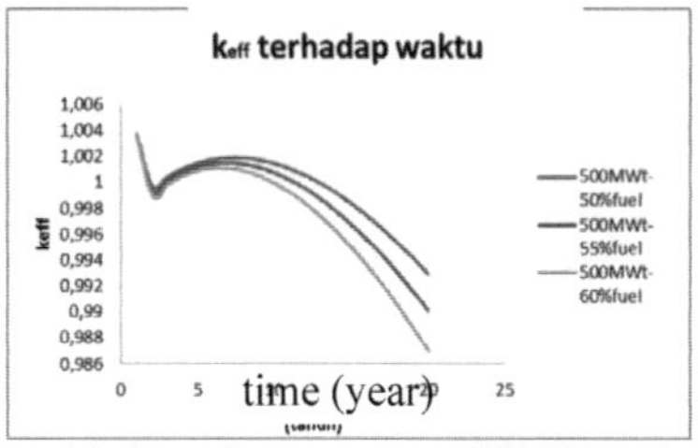

Fig. 10. Curve of k_{eff} in SPINNOR reactor with thermal power output 500MWt and variance fuel fraction 50%, 55% and 60%

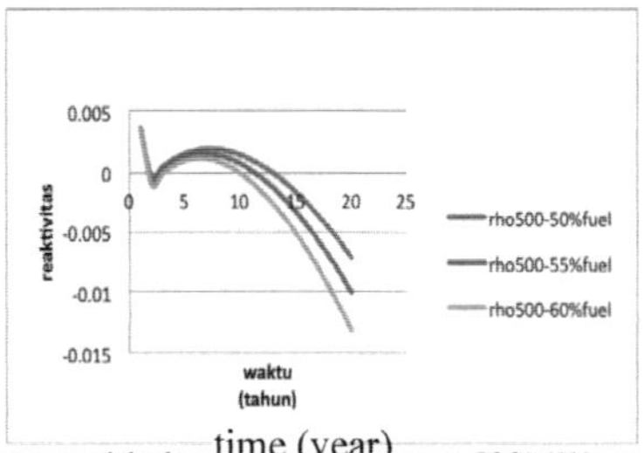

Fig. 11. Curve of reactivity on SPINNOR reactor with ther[illegible]output 500MWt and variance fuel fraction 50%, 55%, and 60%

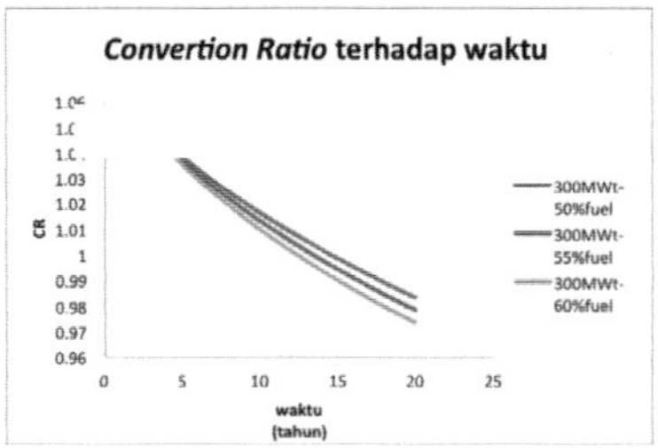

Fig. 12. Curve of *Conversion Ratio* for *Fuel 1* with variance fuel fraction 50%, 55%, and 60% and thermal power output 300MWt

time (year)

K_{eff} and Reactivity Analysis

In figure 6, 8 and 10 show the k_{eff} initial value (year 1) is worth one for each fraction variance of the fuel. For a fraction of the fuel 50%, 55%, and 60% had k_{eff} same value which is equal to 1.00372. The k_{eff} value at the beginning of the process will decline until the 3rd year of operating time for the fuel fraction 50% and 55%, while for 60% fuel fraction k_{eff} value has decreased to the 4th year of operating time. After k_{eff} decreased in the early years of operation, in the next operation will increase the value of k_{eff} and toward a stable state (critical) to the 20th year of the reactor operating time. The critical state of the reactor has been able to achieve 20 years of operation so that it can be said long-life reactor.

Meanwhile, for the value of the reactivity of each fraction can be seen in Figure 7, 9, and 11 where the shape of the graph resembles the graph k_{eff} value. The use of the total thermal output power greater terrace will affect the value of k_{eff} greater Similarly, as in the case of power reactors 300MWt. Of some graphic images k_{eff} and reactivity to the time that has been described above, it can be concluded that in terms of value addition fraction of fuel (fuel), the value of k_{eff} obtained will also rise so that the reactor has a time critical situation long and will extend the operating life of the reactor. This is related to the addition of a fuel fraction, when the fraction or volume of fuel is increased, the amount of fuel will also increase which will cause a number of reactions occurring fissile increased. Increased fissile reaction this would trigger the production of neutrons more with increasing volume fractions. The value of the effective multiplication factor (k_{eff}) is closely related to the number of neutrons in the terrace.

Fuel efficiency can be analyzed using the calculation of excess reactivity of the reactor design. Excess reactivity is a measure of the amount of capacity available to accelerate a nuclear reaction. Value excess reactivity also be used as a parameter whether the reactor can still operate during the first cycle of burn-up fuel. Reactivity value chart in Figure 7, 9, and 11 shows that the results of the calculation of excess reactivity in the fuel fraction 60% and thermal power 300MWt, reactors working optimally for eighteen years of operation. This is because in the 19th operation reactivity value is negative. However, this condition still shows the calculated optimum value of fuel fractions 50-60% in the case. Meanwhile, based on the graph the value of reactivity against time with a thermal power output of 400 and 500 MWT, as in figure (14 and 16) shows that the results of the calculation of excess reactivity to variations fraction of fuel by 50% of the reactor is able to work optimally until the 17th because the reactivity value in year 18 is negative. For a fraction of 55% of the reactor fuel is able to work optimally until the 15th year, and for a fraction of 60% of the reactor fuel is only able to work optimally until the 13th year of reactor operation. However, these conditions illustrates that the reactor is capable of maintaining a critical condition in a long time although it did not reach burn-up period of 20 years.

Analysis of the influence of the value of excess reactivity of a reactor during the burn-up in progress is needed because it will affect the performance of a reactor. Along with the duration of the period of burn-up, the reactor fuel would be less reactive over time. Value excess reactivity greater influence on the increase in reactivity is faster than a nuclear reaction. In normal conditions it must be ensured that the reactivity possessed a reactor must be small enough for the advanced nuclear reactions in the period subsequent burn-up.

In figure 12 show the value of CR (*Convertion Ratio*) to $Fuel_1$ with thermal power output and the variation fraction 300MWt fuel 50%, 55%, and 60% showed CR values above one. On the graph shows that the value of CR in year 1 amounted to 1.054969 and continued to decline until it reaches the value of CR amounted to 1.001312 in the 14th year of reactor operation time. In the 15th until the 20th year CR value reaches a value below one that is approximately 0.99794, of these explanations can be said that the reactor is classified into the reactor Breeding (Breeding Reactor). Breeding Reactor is a reactor that has the ability to produce its own fuel. But for the case of reactor A reactor breeding period is only able to reach a 15-year period of operation.

4. Conclusion

In this study, there are several conclusions, among others:

1. It has been successfully carried out a design/design SPINNOR-cooled fast reactor with a Pb-Bi Thorium Uranium nitride fuel along neutroniknya analysis using SRAC program (Standard Thermal Reactor Analysis Code).
2. Having regard to the critical level of value-level burn-up fuel, the design of a terrace with a fuel fraction 50% Thorium Uranium Nitride most optimal for thermal power reactors.
3. Based on the level of criticality and power variation value, the maximum thermal power that can be produced terrace 150x100cm diameter using a fraction of the fuel input Thorium Uranium Nitride 60%, obtained the maximum thermal power is equal 300MWt.
4. The condition of the reactor design parameters with the variation of the thermal power output of 100, 200, 300MWt, 400MWt, and 500MWt, variations in fuel fraction 50%, 55% and 60%, and the percentage of U-233 by 7%, 7.75%, and 8.5 %. Under these conditions the system is able to maintain the reactor in a critical condition in operation long enough time mecapai 20 years, so that the reactor can be considered as long-life reactor.

5. From the graph k_{eff} and reactivity over time, it can be concluded that in terms of value addition fraction of fuel (fuel), k_{eff} value obtained will also be increased so that the reactor had a long period of critical state and will extend the operating life of the reactor. Because when the fraction or volume of fuel is increased, the amount of fuel will also increase which will cause a number of reactions occurring fissile increased. Increased fissile reaction this would trigger the production of neutrons more with increasing volume fractions.
6. Value excess reactivity indicates that the fast reactor SPINNOR has an optimized design and effective in a state with a fuel fraction 60% for thermal power reactors 300MWt. Because the value of the average excess reactivity to these conditions is 0.200039004% $\Delta k / k$ (< $1).

References

[1] A. Arkundato, Studi Penghambatan Korosi Logam Besi Dalam Lingkungan Pendingin Reaktor Cepat Berbahan Pb dan PbBi Cair Menggunakan Metode Simulasi Dinamika Molekul, Doctoral Dissertation, ITB, 2012.

[2] A.K. Rivai, Z. Su'ud, F. Aziz, Design Study of Modular Fast Reactor with Pb-Bi Coolant and Nitride Fuel, Seminar Tahunan Fisika II, ITB, 2001.

[3] Z. Su'ud, Performansi Keselamatan Reaktor Cepat Berpendingin Timbal dan Campuran Timbal-Bismuth, Prosiding Seminar ke-3 Teknologi dan Keselamatan PLTN dan Fasilitas Nuklir PPTKR-PRSG BATAN, 1995.

[4] Zaki Su'ud: Neutronic Performance Comparation of MOX, Nitride, and Metallic fuel based 25-100 MWe Pb-Bi Cooled Long Life Fast Reactors without on site Refuelling, Progress of Nuclear Energy, **50**, 276-278, 2008.

[5] J.J. Duderstadt, and L.J. Hamilton, Nuclear Reactor Analysis-First edition, John Willey and Sons, New York, 1976.

Fuel Fraction Analysis of 500 MWth Gas Cooled Fast Reactor with Nitride (UN-PuN) Fuel without Refueling

Ratna Dewi Syarifah[1], **Zaki Su'ud, Khairul Basar and Dwi Irwanto**
Nuclear Physics and Biophysics Research Division, Physics Department, Faculty of Mathematics and Natural Science, Bandung Institute of Technology Indonesia, Jalan Ganesha 10 Bandung 40132, INDONESIA

E-mail: syarifah.physics@gmail.com

Abstract. Nuclear Power Plant (NPP) is one of candidates which can support electricity demand in the world. The Generation IV NPP has fourth main objective, i.e. sustainability, economics competitiveness, safety and reliability, and proliferation and physical protection. One of Gen-IV reactor type is Gas Cooled Fast Reactor (GFR). In this study, the analysis of fuel fraction in small GFR with nitride fuel has been done. The calculation was performed by SRAC code, both Pij and CITATION calculation. SRAC2002 system is a code system applicable to analyze the neutronics of variety reactor type. And for the data library used JENDL-3.2. The step of SRAC calculation is fuel pin calculated by Pij calculation until the data homogenized, after it homogenized we calculate core reactor. The variation of fuel fraction is 40% up to 65%. The optimum design of 500MWth GFR without refueling with 10 years burn up time reach when radius F1:F2:F3 = 50cm:30cm:30cm and height F1:F2:F3 =50cm:40cm:30cm, variation percentage Plutonium in F1:F2:F3 = 7%:10%:13%. The optimum fuel fraction is 41% with addition 2% Plutonium weapon grade mix in the fuel. The excess reactivity value in this case 1.848% and the k-eff value is 1.01883. The high burn up reached when the fuel fraction is low. In this study 41% fuel fraction produce faster fissile fuel, so it has highest burn-up level than the other fuel fraction.

1. Introduction

Nuclear Power Plant (NPP) is one of candidate which can support electricity demand in the world. The country that have little fossil fuels or have chosen to use these for feedstock in the petro chemical industry, nuclear power is considered the source of choice for electricity generation. In the United State, nuclear power generates nearly 22,8% of the electricity. In other countries, notably France, the proportion nearly 80% [1]. In 1950s, early prototype reactors has been built, called Generation I of nuclear power. Now, the development of nuclear power is until Generation IV technologies.

In 2002, GIF (Generation-IV International Forum) selected six system from nearly 100 concepts as Generation IV technologies. There are Gas Cooled Fast Reactor (GFR), Lead Cooled Fast Reactor (LFR), Molten Salt Reactor (MSR), Sodium Cooled Fast Reactor (SFR), Supercritical Water Cooled Reactor (SCWR), and Very High Temperature Reactor (VHTR). GIF defined four goals areas to advance nuclear energy, fourth generation. The goals are sustainability, safety and reliability, economic competitiveness, and proliferation resistance and physical protection [2,3].

[1] To whom any correspondence should be addressed.

The GFR system is a high temperature helium cooled fast reactor with a closed fuel cycle. It combines the advantages of fast spectrum systems for long term sustainability of uranium resources and waste minimization (through fuel multiple reprocessing and fission of long-lived actinides), with those of high temperature systems (high thermal cycle efficiency and industrial use of the generated heat) [2,3]. Beforehand, the analyse of GFR with SRAC Code and modified CANDLE burn up scheme has been done [4,5,6,7,8]. In this research, we design 500MWth GFR with nitride (UN-PuN) fuel long life without refuelling.

2. Design Concept and Calculation Methods

The parameter design of the reactor is presented at Table 1. Determination of fuel, cladding, and coolant material in a reactor influence safety and economic factor of the reactor.

Table 1. Parameter design of the reactor

Parameter	Specification
Power	500 MWth
Fuel material	UN-PuN
Cladding material	Stainless Steel
Coolant material	Helium
Fuel volume fraction	40% - 65%
Cladding volume fraction	10%
Coolant volume fraction	30% - 50%
Active core diameter	220 cm
Active height core	110 cm
Reflector	50 cm
Pin pitch	1,45 cm
Reactor life	10 years

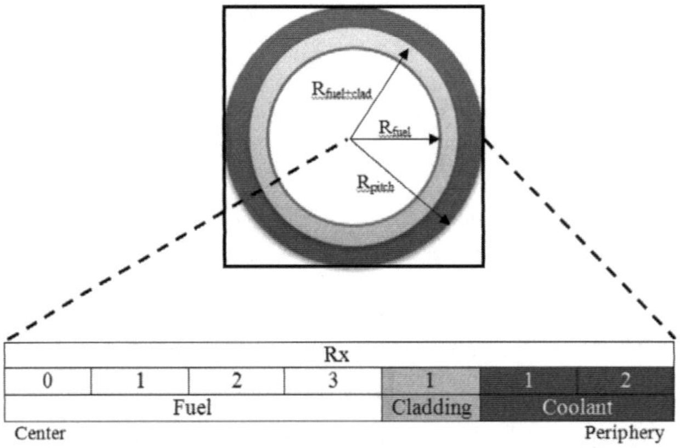

Figure 1. Cell geometry design (square cell) and distribution of region-X

In this study, we use square cell geometry design which presented in Fig. 1. This square cell fuel pin is divided by six regions, the first regions are fuel regions, the fourth region is cladding, and two last regions are coolant regions. A simple division of the region can be seen in fig. 1.

The reactor core use 2-D cylinder geometry design which divided by three regions for fuel. This reactor design will be focused on the optimization of fuel fraction analysis use nitride (UN-PuN) fuel for small GFR. We would like to find, the minimum value of excess reactivity. It means that, the k-eff value is near criticality value (reactor stable). Excess reactivity can be calculated with the equation below.

$$excess.reactivity(\%) = \frac{(k-1)}{k} x100\% \tag{1}$$

Fig. 2 show Half heterogen core configuration with different width in variation fuel. There is three variation fuel, F1, F2 nad F3 which F1 is in the center, F2 in the middle and F3 in periphery, and after that there is a reflector. Percentage F1:F2:F3 = 7%:10%:13%. Table 2 show width of the radius and height in case a and case b which is show in Fig. 2. The calculation is use SRAC2006 code system and data libraries JENDL3.2 [9].

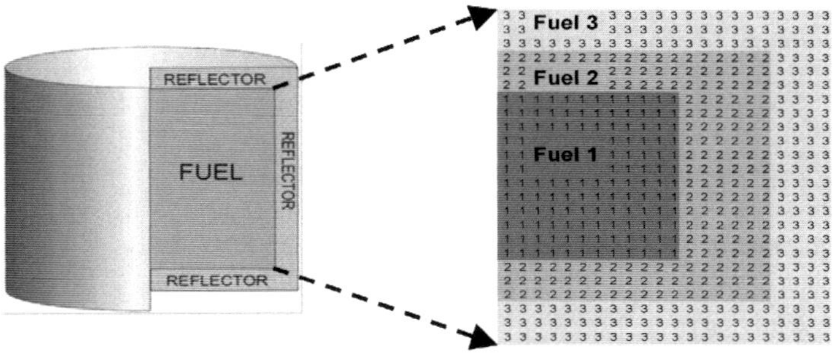

Figure 2. Half heterogeneous core configuration with different width in variation fuel

Table 2. Width of the radius and height in two case (case a and case b)

Parameter	Fuel type	Case a	Case b
Radius (cm)	Fuel 1	60	50
	Fuel 2	30	30
	Fuel 3	20	30
Height (cm)	Fuel 1	60	50
	Fuel 2	40	40
	Fuel 3	20	30

3. Results and Discussion

In the first step calculation, we calculate k-eff value for case a and case b (see Table 2). Fig. 4 show k-eff value of case comparation with three variation fuel. The k-eff value increase when the reactor use radius 50 cm for F1, 30 cm for F2 and 30 cm for F3. And after burn up time 2 years, k-eff value is stable. It means that when it decrease F1 volume, otomaticaly the volume of F2 and F3 increase, it make plus Plutonium in core reactor, so the k-eff value is increase.

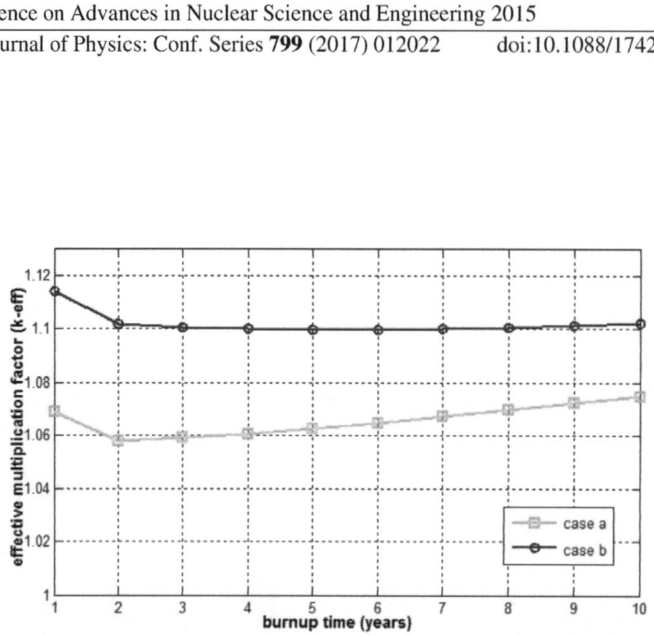

Figure 4. K-eff value of case comparation with three variation fuel F1:F2:F3=7%:10%:13%

Fig. 5 show k-eff value of fuel fraction comparation with three variation fuel. The fuel fraction variation is from 40% until 60% fuel fraction. Fig. 6 show the fuel fraction variation does not change the trend of the graph. The graph trend is similar, but it influence the k-eff value. When the fuel fraction is 40%, the k-eff value is subcritic. And when the fuel fraction is 45%, the k-eff value is around 1,03. It means that the excess reactivity still high. In this research, the objective of this research is to design the reactor which have 10 years burn up time and low excess reactivity value (less than 2%), so we must optimize the design.

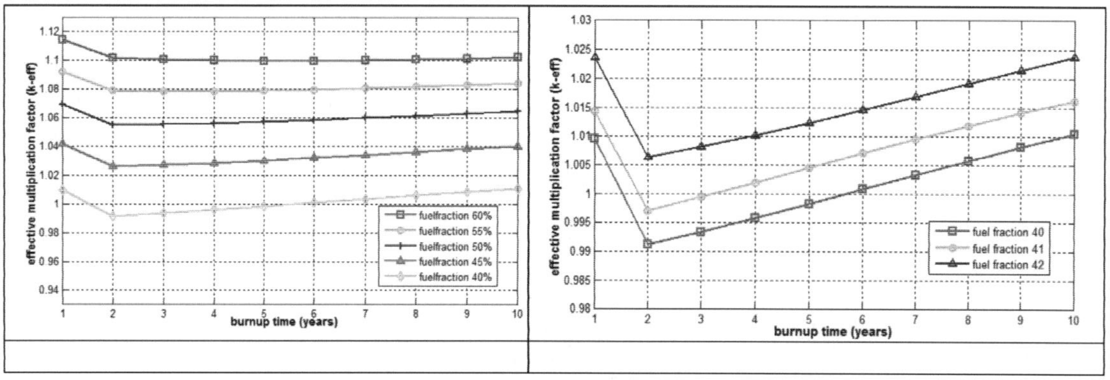

Figure 5 (a) K-eff value of fuel fraction comparation with three variation fuel F1:F2:F3=7%:10%:13% from fuel fraction 40% up to 60%, (b) K-eff value of fuel fraction 40%, 41% and 42%

Fig. 5 (b) show k-eff value of fuel fraction 40%, 41% and 42%. Due to the objective is to find excess reactivity value less than 2%, the fuel fraction calculation calculate more detail than before. From Fig. 6 show that excess reactivity value of fuel fraction 42% more than 2% or around 2,4%. Whereas, k-eff value in first and second burn up time for fuel fraction 41% is still subcritic. So, it take a weapon grade (Pu-239 pure) to increase the k-eff value. The philosophy take Pu weapon grade in the reactor is to decrease Pu-239 in the world, so it can't be used in nuclear weapon.

Fig. 4 and 5 show that there are always a peaking graph in the first until second years burn up time. The peaking value because the percentage of Pu-241 in the Plutonium isotope from waste PWR. Table 3 show plutonium percentage of total Plutonium in core from waste PWR. Percentage of Pu-241 is 11,4%, it is to much, because it absorb the neutron more than Pu-239. Because the cross section absorbtion value of Pu-241 bigger than cross section absorbtion value of Pu-239. So, the k-eff value is decrease greatly in the first step burn up.

Table 3. Plutonium percentage of total Plutonium from waste PWR (Waltar, 1981)

Isotope	PWR (U-fueled) Burn up 33 MWd/ton
Pu-238	1.8%
Pu-239	58.7%
Pu-240	24.2%
Pu-241	11.4%
Pu-242	3.9%

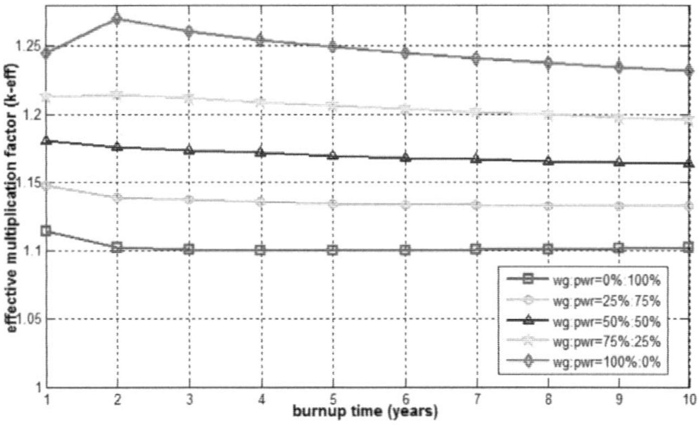

Figure 6. K-eff value of comparation Plutonium weapon grade (Pu-239 pure) and Plutonium from waste PWR fuel

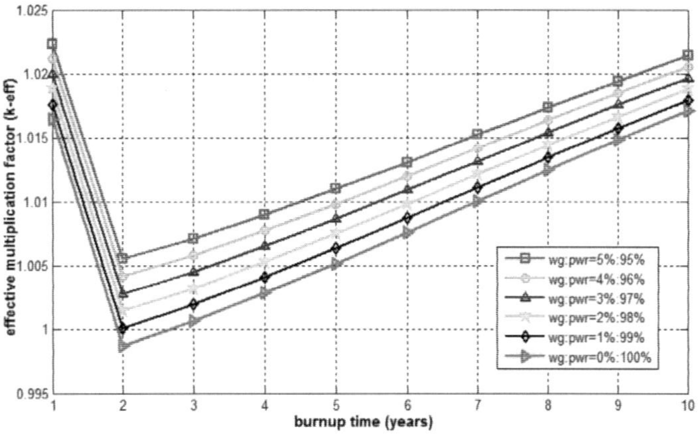

Figure 7. K-eff value of addition Pu weapon grade from 0% up to 5%

Due to, the k-eff value is decrease greatly, it take Plutonium weapon grade (pure Pu-239) mix into fuel. Fig. 6 show k-eff value of comparation Plutonium weapon grade and Plutonium from waste PWR. The variation is from 0%, 25%, 50%, 75% and 100%. Fig. 6 shows that the addition of Plutonium weapon grade change the trend of the graph in the first and second burn up time. Because Pu weapon grade can change the trend of the graph, we take Pu weapon grade in the calculation.

Figure 7 show k-eff value of addition Plutonium weapon grade from 0% up to 5%. From Figure 7, the optimum k-eff value when addition 2% of Pu weapon grade. The maximum excess reactivity value in this case 1.848% (less than 2%). And the maximum of k-eff value is 1.01883.

4. Conclusion

The optimum design of 500MWth GFR without refueling with 10 years burn up time reach when radius F1:F2:F3 = 50cm:30cm:30cm and height F1:F2:F3 =50cm:40cm:30cm, variation percentage Plutonium in F1:F2:F3 = 7%:10%:13%. The optimum fuel fraction is 41% with addition 2% Plutonium weapon grade mix in the fuel. The excess reactivity value in this case 1,848% and the k-eff value is 1,01883. The high burn up reached when the fuel fraction is low. In this study 41% fuel fraction produce faster fissile fuel, so it has highest burn-up level than the other fuel fraction.

References

[1] Lamarsh J R 1983 *Introduction to Nuclear Engineering* (Addison-Wesley Publishing Company Inc Massachusetts) p 1-2

[2] GIF (The Generation IV International Forum) 2002 *A technology Roadmap for Generation IV Nuclear Energy System* (U.S DOE Nuclear Energy Research Advisory Committee) p 3-8

[3] GIF (The Generation IV International Forum) 2014 *Technology Roadmap Update for Generation IV Nuclear Energy System* (the OECD Nuclear Energy Agency) p 3-8

[4] Menik A Zaki S Fiber M Abdul W Khairurrijal Idam A Aziz F and Hiroshi S 2013 Optimization of small long life gas cooled fast reactors with natural Uranium as fuel cycle input *Applied Mechanics and Materials* **261-262** pp 307-311

[5] Zaki S and Hiroshi S 2013 The prospect of gas cooled fast reactors for long life reactors with natural uranium as fuel cycle input *Annals of Nuclear Energy* **54** pp 58-66

[6] Fiber M Zaki S Abdul W Khairul B Menik A and Hiroshi S 2013 Application of Modified CANDLE burn up to very small long life gas cooled fast reactor *Advanced Material Research* **772** pp 501-506

[7] Fiber M Menik A Zaki S Abdul W Khairul B Ferhat A Sidik P and Hiroshi S 2014 Conceptual design study on very small long-life gas cooled fast reactor using metallic natural Uranium-Zr as fuel cycle input *AIP Conference Proceedings* **1584** pp 105-108

[8] Fiber M Zaki S Abdul W Khairul B Menik A and Hiroshi S 2014 Power flattening on modified CANDLE small long life gas-cooled fast reactor *AIP Conference Proceedings* **1615** pp 47-50

[9] Okumura K 2002 *SRAC version 2002* (Japan Atomic Energy Research Institute) p 1-28

Space and Time Distribution of Pu Isotopes inside The First Experimental Fuel Pin Designed for PWR and Manufactured in Indonesia

Suwardi[1]**, J. Setiawan**[1]**, J. Susilo**[2]

[1]Center for Nuclear Fuel Technology-BATAN,Puspiptek Bld.20, Tangerang-Selatan,15310, Indonesia
[1]Center for Nuclear Safety Technology-BATAN,Puspiptek Bld.60, Tangerang-Selatan,15310, Indonesia

*Email: suwardi@batan.go.id

Abstract. The first short fuel pin containing natural UO_2 pellet in Zry4 cladding has been prepared and planned to be tested in power ramp irradiation. An irradiation test should be designed to allow an experiment can be performed safely and giving maximum results of many performance aspects of design and manufacturing. Performance analysis to the fuel specimen shows that the specimen is not match to be used for power ramp testing. Enlargement by 0.20 mm of pellet diameter has been proposed. The present work is evaluation of modified design for important aspect of isotopic Pu distribution during irradiation test, because generated Pu isotopes in natural UO2 fuel, contribute more power relative to the contribution by enriched UO2 fuel. The axial profile of neutrons flux have been chosen from both experimental measurement and model calculation. The parameters of ramp power has been obtained from statistical experiment data. A simplified and typical base-load commercial PHWR profile of LHR history has been chosen, to determine the minimum irradiation time before ramp test can be performed. The data design and Mat pro XI materials properties models have been chosen. The axial profile of neutrons flux has been accommodated by 5 slices of discrete pin. The Pu distribution of slice-4 with highest power rate has been chosen to be evaluated. The radial discretion of pellet and cladding and numerical parameter have been used the default best practice of TU. The results shows that Pu 239 increased rapidly. The maximum burn up of slice 4 at upper the median slice, it reached nearly 90% of maximum value at about 6000 h with peak of 0.8%a Pu/HM at 22000 h, which is higher than initial U 235. Each 240, 241 and 240 Pu grows slower and ends up to 0.4, 0.2 and 0.18 % respectively. This results can be used for verification of other aspect of fuel behavior in the modeling results and also can be used as guide and comparison to the future post irradiation examination for Pu isotopes distribution.

1. Introduction

Experimental Fuel Elements Installation (EFEI) re-vitalization that began be built late 20th century needs to be improved and with future goals change of EFEI from only laboratory for CIRENE BHWR (Boiling Heavy Water Reactor) fuel technology Power Reactor Fuel type of Cirene, to gradually improved its ability of preparation of various specimens of fuel elements from different types of reactors, PHWR (Pressurized Heavy Water Reactor), especialy CANDU then in the end also will be one the first factory-owned BATAN to produce Power Reactor Fuel type of CANDU, if Candu type nuclear power plants have been later. Preparation direction has been initiated, and at times can realisa thereof can be started, if desired [1].

The first short fuel pin containing natural UO_2 pellet in Zry4 (Zircaloy-4) cladding has been prepared at the CNFT. The PRTF installed inside RSG-GAS (*Reaktor Serba Guna G.A.Siwabessy*) multi purpose 30 MW reactor has been revitalized and is ready for irradiation of fuel pin speciment. An irradiation test should be designed to allow the experiment can be performed safely and giving maximum results of many performance aspects of design and manufacturing. The present work is evaluation of important isotopic Pu generation-depletion and distribution during irradiation test, as Pu generated Pu during irradiation in natural UO2 fuel gives more power fraction than fuel containing enriched UO2. The constraint of axial profile of neutrons flux and the maximum generated integral power have been determined from both experimental measurement and modeling calculation. The contraint of maximum ramp power has been obtained from measured experiment data [2]. As natural uranium is used for fuel pellets, a simplified and typical base-load commercial PHWR profile of LHR (Linear Heat Rate) history has been chosen in determining the minimum irradiation time before ramp test can be performed. The obtained design and manufactured fuel pin data and measured and model material properties data and behavior and component have been used as input.

The natural-UO2 fuel geometry is in between former design of PWR fuel and new design of PHWR fuel. The plan of irradiation test will be at PWR coolant temperature & pressure and thermal neutron flux of PWR.

The chosen model takes into account mat-pro XI [3] materials properties models, model for isotopes generation and depletion but ignores the period of power shut-down and restart-up. Typical high power rate has been chosen to be evaluated. The axial profile of neutrons flux has been accommadated by 5 slices of discret pin. The radial discretizatioon of pellet and cladding and numerical parameter have been used the default best paractice of TransUranus [4] code used. The code has been veried by diference theoretical and computation modeling and experiment data [5] The results of time distribution of Pu 239 increased rapidly. The maximum burnup of slice 4 at upper the median slice, it reached nearly 90% of maximum value at about 6000 h with peak of 0.8%a Pu/HM at 22000 h, wich is higher than initial U 235. Each 240, 241 and 240 Pu grows slower and ends up to 0.4, 0.2 and 0.18 % respectively. The axial distribution of Pu isotopes are proportionaly to the flux profiles of thermal neutron. This results can be used for verification of other aspect of fuel behavior in the modeling results and also can be used as guide and comparison to the future post irradiation examination for Pu isotopes distribution.

2. Theory / Calculation

Fig. 0 illustrates the chain of nuclear fission and decay of Pu isotopes [6]. The most common isotope formed in a typical nuclear reactor is the fissile Pu-239 isotope, formed by neutron capture from U-238 (followed by beta decay), and which yields much the same energy as the fission of U-235. Well over half of the plutonium created in the reactor core is 'burned' *in situ* and is responsible for about one third of the total heat output of a light water reactor (LWR). Of the rest, about one sixth through neutron capture becomes Pu-240 (and Pu-241). The approximately 1.15% of plutonium in the spent fuel removed from a commercial LWR power reactor (burn-up of 42 GWd/t) consists of about 53% Pu-239, 25% Pu-240, 15% Pu-241, 5% Pu-242 and 2% of Pu-238, which is the main source of heat and radioactivity [7]

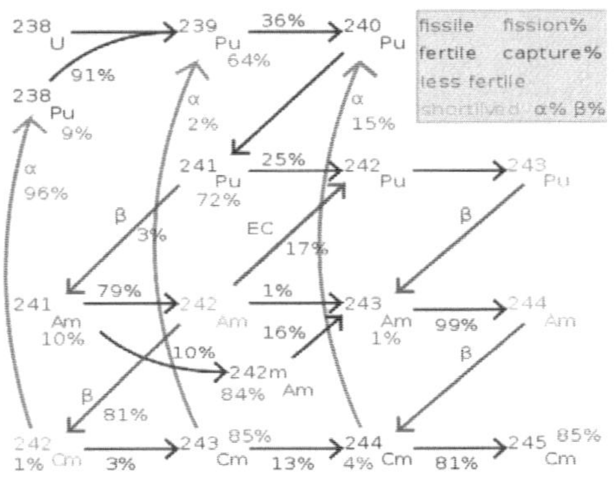

Fig. 0. Nuclear reaction chain of Pu isotopes during irradiation of natural UO2 pellet[6].

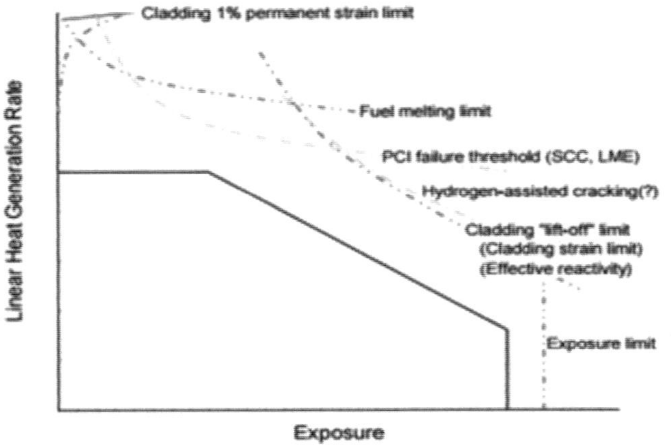

Fig.1. LHR profile chosen under LHR envelop and related its individual constraints [11]

3. Experimental Method

The natural UO2 fuel geometry is in between former design of PWR fuel [11] and new design of PHWR fuel of larger pellet and thinner fuel to cladding gap. The plan of irradiation test will be at PWR coolant temperature & pressure and thermal neutron flux of PWR. The main purpose of the irradiation test in the PRTF is to obtain the limite of power ramp can be applied to the fuel pin. The power ramp parameter to be considered for the testing must be determined according to the design and fabrication. Since the design is chosen need to be compatible to both the existing PTRF which are capability and duty limites and capability or mastering of fabrication of pin technology, while the main required fuel performance for the first NPP in Indonesia is: minimum fuel assemby discharge burnup 50 MWd/kgHE in case the reactor type is PWR. The comparative natural uranium fuel or HWR fuel burnup is about 8 MWd/kgHE. The problem is solved in the platfoem of Transuranus integral thermal mechanical behavior of water cooled reactor.

The fuel pin design is presented in Table-1, which is sorted form originale of EdyS publication [8]. The LHR history is chosen a typical simple one under envelop described by Hagrman [9]. Despite the accompany lot properties of cladding exist, the prediction use the standard MatPro. Both properties of pellet and cladding are modelled according to MatPro xi, although the accompany lot properties of cladding exist. Since the existing accompanied properties of cladding lot is fulfil the ASTM B850 [10].

The axial profile of neutrons flux have been chosen from both experimental measurement and model calculation. The parameter of ramp power has been obtained from statistical experiment data. As natural uranium is used for fuel pellets, a simplified and typical base-load commercial PHWR profile of LHR history has been chosen, to determine the minimum irradiation time before ramp test can be performed. The obtained design and manufactured fuel pin data and measured and model material properties data and behavior and component have been used as input. The chosen model takes into account MatPro XI materials properties models, HWR model for isotope generation and depletion but ignores the period of power shut-down and restart-up. The axial profile of neutrons flux has been accommodated by 5 slices of discrete pin. The slice-4 with highest power rate has been chosen to be evaluated. The radial discretion of pellet and cladding and numerical parameter have been used the default best practice of TU code used

The LHR history has been chosen under LHR limite by combination of different constraint from begining of life (BOL) up to end of line (EOL) [11]

1) Cladding 1% permanent strain limit at BOL (Begining of Live); 2) Fuel melting limit; 3)PCI failure threshold; 4)Hydrogen assisted cracking ; 5)Cladding lift-off limit / Cladding strain limit / Effective reactivity 6) Exposure limit While the cooling temperature and pressure is typical base load of commercial irradiation. The important input of initial and boundary condition is presented by Table.2.

4. Results and Discussion

The results of time distribution of Pu 239 increased rapidly. The maximum burn up of slice 4 at upper the median slice, it reached nearly 90% of maximum value at about 6000 h with peak of 0.8%a Pu/HM at 22000 h, which is higher than initial U 235. Each 240, 241 and 240 Pu grows slower and ends up to 0.4, 0.2 and 0.18 % respectively. The axial distribution of Pu isotopes are proportionally to the flux profile of thermal neutron. This results can be used for verification of other aspect of fuel behavior in the modeling results and also can be used as guide and comparison to the future post irradiation examination for Pu isotopes distribution.

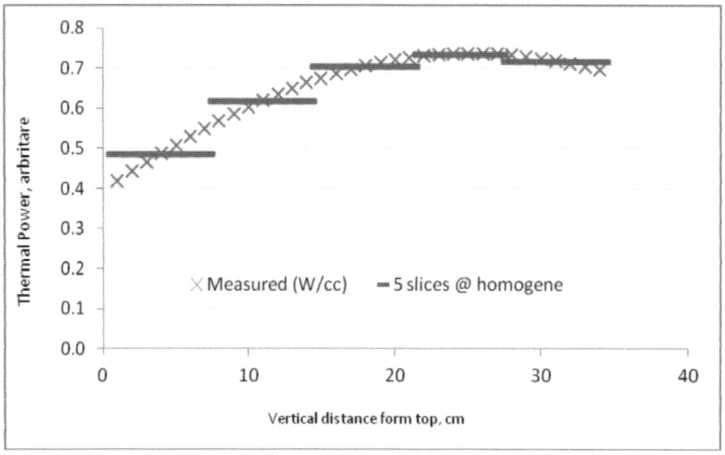

Fig. 2a. Axial (absis) distribution of power density computed form neutronic (x blue) and model of 5 slices (--red)

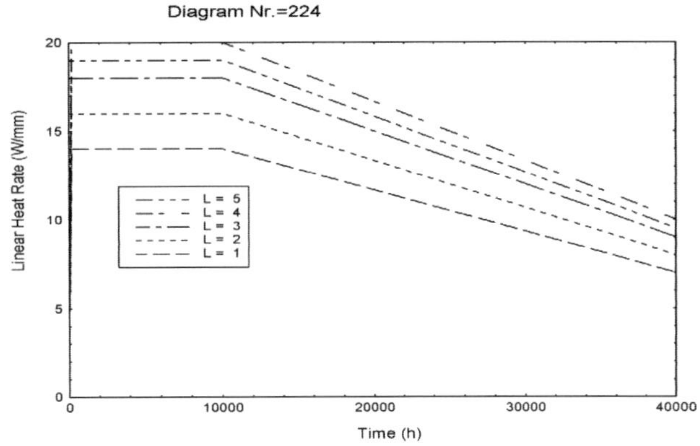

Fig.2b. Chosen input LHR history of different axial positions of fuel pin

Fig. 3 shows the temporal distribution of 5 isotopes and its total, of radially averaged slice-4 whichis highest power of axial position, from start-up to 20000 hr. The Pu 239 the highest concentration of Pu Isotope first grows rapidly up to about 5000 hr then grows more slowly accomponied to the growth of Pu240, followed by Pu241 and Pu242 consecutively. The fraction of Pu239 to the total Pu reach its peak at about 5000 hr, or in this case 6000 hr, which is 0.55, then slowly decreasing. This peak is chosen for Pu-239 production from natural UO2.

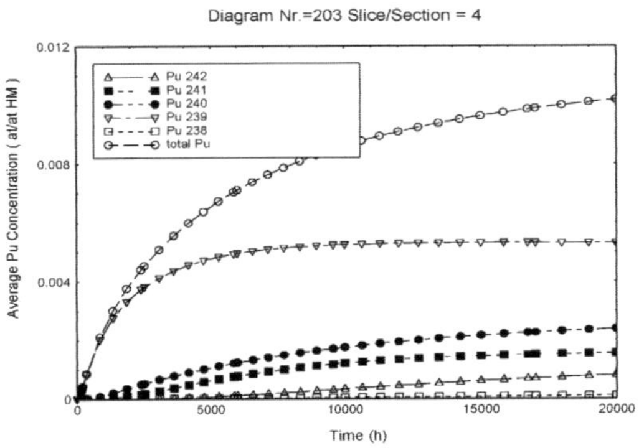

Fig.3. Generation and depletion of Different Pu Isotopes

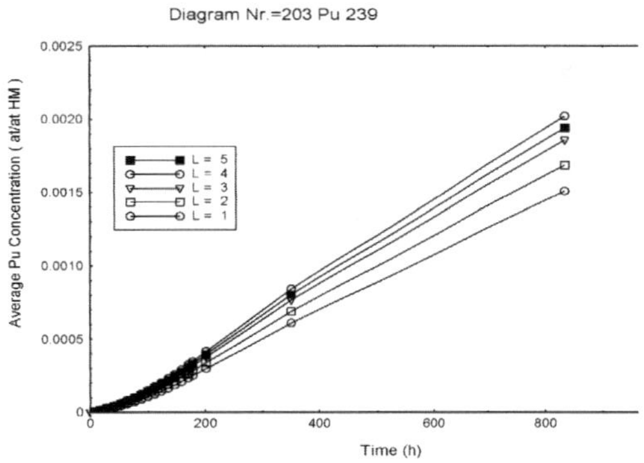

Fig.4. Time distribution of Different Pu 239 for different axial positions of fuel pin

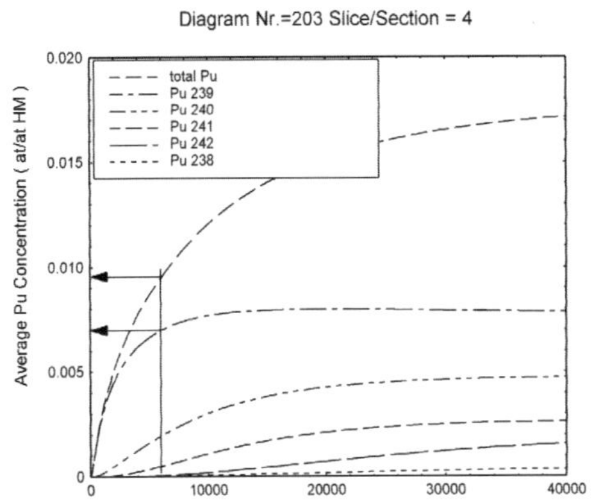

Fig. 3 (b) Zoom of Fig. 3Pu_i in slices-4 up to 40.000 h {40000 0.017153 Pu-total}

Table-1. Fuel pin Data [12]

Fuel Pin Parameter	Value	Unit
Pellet:		
Material	Nat. UO_2	-
Length	12	mm
Outside diameter	9.55	mm
Density	95	% ThD
Cladding:		
Material	Zry-4	-
Thickness	1	mm
Number of Pellet per pin	20	-
Rod:		
Length	244	mm
Outside diameter	13,2	mm

IOP Conf. Series: Journal of Physics: Conf. Series **799** (2017) 012023 doi:10.1088/1742-6596/799/1/012023

Fig. 4 shows the evolution of radially averaged of slices for Pu239. Slice-4 produces highest Pu239, as it is the highest generating power.

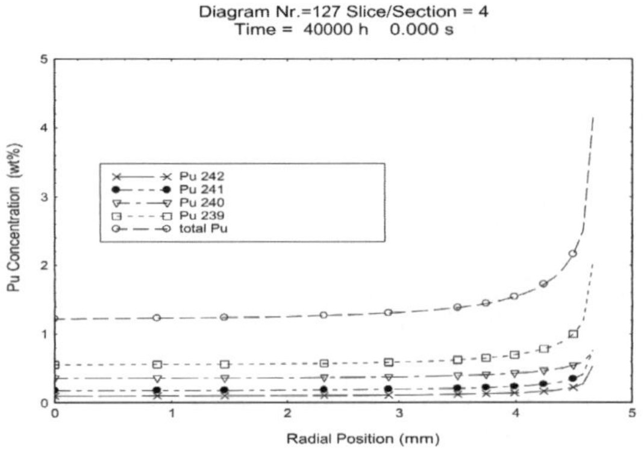

Fig.5. Radial distribution of Different Pu Isotopes at the end of irradiation

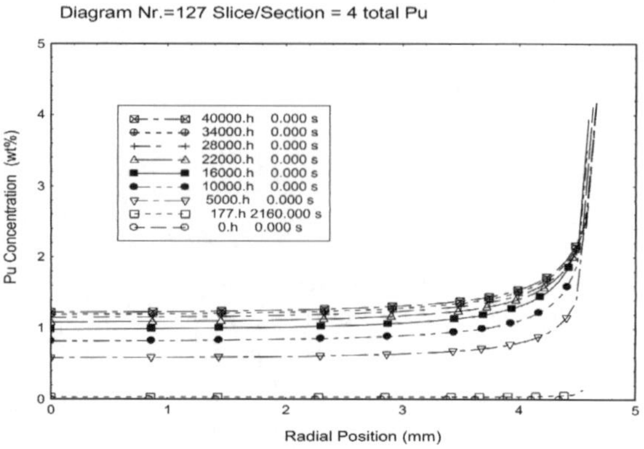

Fig.6. Radial distribution of total Pu Isotopes at different irradiation times

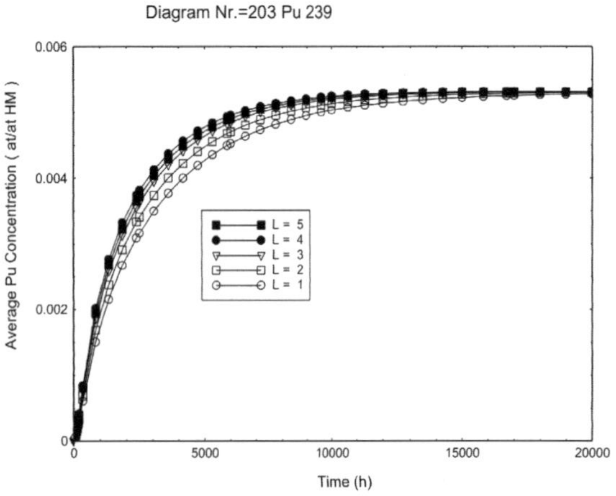

Fig.7. Evolution of average concentration of Pu 239 in different slices up tp 20.000 h

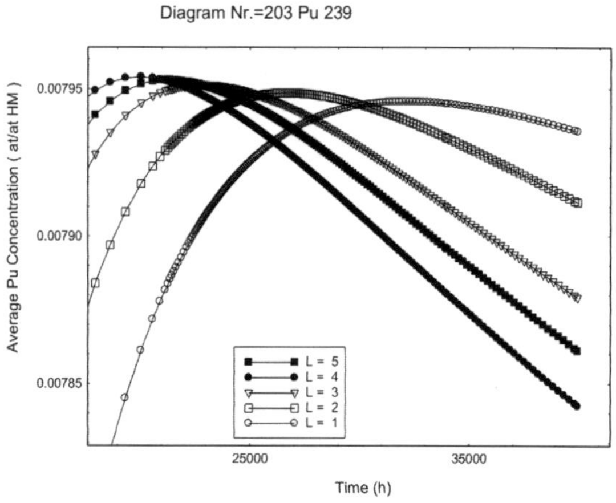

Fig.8. Zoom Fig 7. Arround 25000 hr, the volution of average concentration of Pu 239 in different slices during (20 – 40) 000 h

Fig. 5 shows the radial distribution of Pu Isotopes accentuating with time irradiation or burnup for all isotope at 10-15% of outer zone of pellet as shown by Fig 5 for BOL. Fig. 6 shows this is not only at BOL but aaccentuating all irradiation time. It related to penetration limit of epithermal neutron, and termal flux distribution. The data can be used for minimizing and optimizing the post irradiation examination.

Table-2. Important initial boundary data

No. of section or slice		1	2	3	4	5
Axial coordinate	(mm)	31.96	95.88	159.80	223.72	257.65
Relative position	(%)	12.31	36.93	61.55	86.18	99.24
Linear rating	(kW/m)	0.00	0.00	0.00	0.00	0.00
Fuel temperature	(C)	20.00	20.00	20.00	20.00	20.00
Coolant temperature	(C)	20.00	20.00	20.00	20.00	20.00
Steam content	(%)	0.00	0.00	0.00	0.00	0.00
Heat transfer coeff.	(W/m**2*K)	0.11E+05	0.11E+05	0.11E+05	0.11E+05	0.11E+05
Heat transfer coeff.tot	(W/m**2*K)	0.11E+05	0.11E+05	0.11E+05	0.11E+05	0.11E+05
Melt fraction fuel	(%)	0.00	0.00	0.00	0.00	0.00
Melt fraction clad	(%)	0.00	0.00	0.00	0.00	0.00
Fuel inner radius	(mm)	0.000	0.000	0.000	0.000	0.000
Fuel outer radius	(mm)	4.515	4.515	4.515	4.515	4.515
Thickness HBS, transition	(mm)	0.000	0.000	0.000	0.000	0.000
Thickness HBS, fully develop.	(mm)	0.000	0.000	0.000	0.000	0.000
Clad inner radius	(mm)	4.665	4.665	4.665	4.665	4.665
Clad outer radius	(mm)	5.375	5.375	5.375	5.375	5.375
Total clad outer radius	(mm)	5.375	5.375	5.375	5.375	5.375
Outer clad corrosion layer	(um)	0.00	0.00	0.00	0.00	0.00
Relative radial clad deformation:						
inner radius	(%)	0.00	0.00	0.00	0.00	0.00
outer radius	(%)	0.00	0.00	0.00	0.00	0.00
Average strains in cladding:						
effective creep strain	(%)	0.00	0.00	0.00	0.00	0.00
effective plastic strain	(%)	0.00	0.00	0.00	0.00	0.00
permanent tangent.strain	(%)	0.00	0.00	0.00	0.00	0.00
Cavity pressure	(MPa)	0.10	0.10	0.10	0.10	0.10
Clad inner loading	(MPa)	0.10	0.10	0.10	0.10	0.10
Outer pressure	(MPa)	0.10	0.10	0.10	0.10	0.10
Fluence	(neutrons/(cm**2))	0.00E+00	0.00E+00	0.00E+00	0.00E+00	0.00E+00
Gap conductance	(W/m**2*K)	894.	894.	894.	894.	894.
Gap size (radial)	(um)	149.64	149.64	149.64	149.64	149.64
Centre grain size	(um)	10.00	10.00	10.00	10.00	10.00
Burn-up	(MWd/tU)	0.	0.	0.	0.	0.
Fission gas release	(%)	0.00	0.00	0.00	0.00	0.00
He production	(umol/mm)	0.00E+00	0.00E+00	0.00E+00	0.00E+00	0.00E+00
He release	(%)	0.00	0.00	0.00	0.00	0.00

Fig. 7 shows average Pu239 concentration of each 5 slices. It shows the slice-4 is the highest, but also the first both increasing and in Fig. 8 decreasing, as its U238 is first exhausted because the original fissile is homogeneously distributed, but the LHR and burn-up are the highest.

Table 3. Typical concentration of Pu isotopes as % of fissile content at different reactor type and burn-up

Reactor type	Mean fuel burn-up (MW d/t)	Percentage of Pu isotopes at discharge					Fissile content %
		Pu-238	Pu-239	Pu-240	Pu-241	Pu-242	
PWR	33000	1.3	56.6	23.2	13.9	4.7	70.5
	43000	2.0	52.5	24.1	14.7	6.2	67.2
	53000	2.7	50.4	24.1	15.2	7.1	65.6
BWR	27500	2.6	59.8	23.7	10.6	3.3	70.4
	30400	N/A	56.8	23.8	14.3	5.1	71.1
CANDU	7500	N/A	66.6	26.6	5.3	1.5	71.9
AGR	18000	0.6	53.7	30.8	9.9	5.0	63.6
Magnox	3000	0.1	80	16.9	2.7	0.3	82.7
	5000	N/A	68.5	25.0	5.3	1.2	73.8

Pu240/Pu239 increases along irradiation time. Higher content in Pu240 is better for proliferation resistance, since difficulty in Pu239 purification. After 20000 hr the Pu 239 content decreases continuously while other Pu isotopes increases. The maximum Pu239 concentration 20000 hr reach its maximum 7.95% at/at of HM

Fig. 8 shows the temperature of outer cladding and bulk coolant of first slice, as well as inner cladding and outer surface of pellet. At about 18000 hr both surface temperature of pellet and inner cladding increases sharply then drops slowly, but both temperature of outer cladding and bulk coolant slightly decreasing. The inner temperature of cladding reached 1305 °C, but the outer one is still 250 °C. This temperature is still under its melting point, but the related cladding creep need to be considered. Yes the cladding creep is positive and the lift-off is out of limit.

5. Conclusion

CNFT has prepared a short pin as fuel pin prototype based on natural UO_2 pellet. An irradiation test will performed in the PRTF. A simulation of irradiation test has been performed for to support testing license. The spatial and temporal distribution of generated and depleted Pu 239, 240, 241 and 242 and 238 isotopes has been analyzed as Pu239 produced and partially fissioned resulting important additional power.

Axial distribution of radially averaged concentration of Pu isotopes at certain burn up the axial power.

Radial distribution of Pu isotopes by slice-4 for different irradiation time has been visualized. All radial distribution has typical having peak at the surface of pellet proportionally to radius position of U burn-up. All peak of different isotopes tend to accentuates as irradiation time or burn-up accumulated.

Temporal distribution of different Pu isotopes is characterized by highest and sunner appearance of Pu-239 concentration. The ratio of Pu239 to total Pu isotope reach its peak at 6000 hr, corresponding to optimum irradiation time for Pu -239 production.

These visualization of radial – axial and temporal distribution of different Pu isotopes which are calculated for distribution of power, temperature, mechanical properties and eventually to chemistry behavior of cladding thermal and mechanical behavior of fuel pin can be used to benchmarking of post irradiation examination.

Acknowldgment

The authors want to thank to Mr. Basak the Coordinator of IAEA CRP-1127 who made the author realize their research plan and also to their partners of Institute of Transuranium Element - Joint Research Centre of European Commission for licensing of the TransUranus code written by K. Lasmann used in this work. The authors would address acknowledgment to Mr. J. Susilo Mr. J. Iman and Mr, W. Dewayatna from BATAN for supporting neutronic data

References

[1] Suwardi, 2013, Parametric Study on irradiation test of CNFT fuel prototype, Proc. WWER Fuel Performance Modeling and Experiment Support, Sandanski Bulgaria, 2013
[2] Anonym, 1983, Manufacturing CIRENE fuel by ENEA, Comb. IEEC, Nuclear Fuel Manufacturing Division, ITALY.
[3] H.Sudirdjo et al. 2005, Revitalization of Power Ramp Test Facility of RSG-GAS Reactor, Proc. Seminar Research Reactor Technology and Application, 2005, BATAN.
[4] Anonym, 2011, Transuranus Handbook, European Commission, Joint Research Centre-JRC, ITU 2011.
[5] P. VAN UFFELEN and M. SUZUKI, 2011, Chapter 71 "Fuel Performance Modeling and Simulations: Oxide Fuel Performance Modeling and Simulation" in Comprehensive Nuclear Materials, edited by R. Konings, accepted for publication in J. Nucl. Mat., 2011.
[6] A. SCHUBERT, P. VAN UFFELEN, J. van de LAAR, C. T. WALKER, W. HAECK, 2008, Journal of Nuclear Materials,Vol. 376 (2008) 1-10.
[7] P. VAN UFFELEN, Cs. GYÖRI, A. SCHUBERT, J. van de LAAR, Z. HÓZER, G. SPYKMAN, 2008, Journal of Nuclear Materials, Vol. 383 (2008) 137-143.

[8] T.Taryo and T.M. Sembiring, 2014, Analysis of Neutron Flux Distribution in RSG-GAS Reactor with U-Mo Fuels, Center for Development of Research Reactor Technology-BATAN, Tangerang, Indonesia, 2014.
[9] B.Arbie, 1987, Bakri ARBIE and Sutaryo SUPADI, Status Report on the Indonesian Multipurpose Reactor RSG-GAS, BATAN, Indonesia, 1987.
[10] J. Iman, 2015, privat communication, RSG-GAS, Serpong, Indonesia.
[11] J. Susilo, 2015, privat communication, Center for Nuclear Technology and Reactor Safety Serpong, Indonesia.
[12] L.P. Hong and T.M. Sembiring, Nuclear Engineering and Design 240 (2010)1433-1442.
[13] Edy Sulistyono and Tri Yulianto. J.Tek.Bhn.Nuk vol 9, No.2 (2013), pp 55-66
[14] K. Lassmann, TRANSURANUS: a fuel rod analysis code ready for use, J. Nucl. Mater., 188 (1992), pp. 295–302
[15] B. D. Murphy, Prediction of the Isotopic Composition of UO2 Fuel from a BWR: Analysis of the DU1 Sample from the Dodewaard Reactor, Computational Physics and Engineering Division, OAK RIDGE NATIONAL LABORATORY, 1998
[16] G. Kyriazidis, P Marimbeau, J. Cl Lefebvre. IMPACT OF IRRADIATION HISTORY ON INVENTORY PREDICTION DATA FOR PWR FUEL. PHYSOR 1994 - Re-actor Physics Faces 21st Century , Apr 1994, Knoxville - Tennessee, United States. Actes de la conference. <cea-01122087>
[17] L. Holta, b, , , U. Rohdea, M. Seidlc, A. Schubertd, P. Van Uffelend, R. Macián-Juanb Development of a general coupling interface for the fuel performance code TRANSURANUS – Tested with the reactor dynamics code DYN3D, Annals of Nuclear Energy, Volume 84, October 2015, Pages 73–85
[18] B. D. Murphy, Prediction of the Isotopic Composition of UO2 Fuel from a BWR: Analysis of the DU1 Sample from the Dodewaard Reactor, Computational Physics and Engineering Division, OAK RIDGE NATIONAL LABORATORY, 1998
[19] S. Azzaoui, SCALE-6 fuel depletion analyses:Application to the ARIANE program, EXTERNAL REPORT OF THE BELGIAN NUCLEAR RESEARCH CENTRE, August, 2010, SCK.CEN, Boeretang 200, Belgium

Radii and Density Calculations of ^{209}Bi by Using *Skyrme-Hartree-Fock* Method

Yacobus Yulianto[1, a], **Zaki Su'ud**[1, b]
[1]Nuclear and Biophysics Research Division, Physics Department,
Faculty of Mathematics and Natural Science, Institut Teknologi Bandung,
Jl. Ganesha 10 Bandung, 40132, Indonesia.

E-mail: [a]yacyulianto@gmail.com, [b]zakisuud@gmail.com

Abstract. In this study, the densities, the neutron skin thickness, and the rms radii for both proton and neutron of ^{209}Bi are calculated using *Hartree-Fock* method with *Skyrme* set parameters, especially *SLy4*, *SkM**, Z_σ, and *SIII* set parameters. All obtained results are in good agreement with the related experiment and the other researcher results.

1. Introduction

One of the basic quantities to describe nuclear structure is the nuclear density distribution, which can give detail information on the nuclei internal structure which related to proton and neutron wave function. The other quantities are the nuclear charge radii which represent the most useful observables information in nuclear structure analysis [1]. These observables, which can be determined from the form factors, provide information about the nuclear shape. To study about nuclear structure, especially for nuclei far from stability line, *Hartree-Fock* method combined with *Skyrme* interaction provides a very successful microscopic model to describe many nuclear properties, such as nuclear ground states, collective motions, fission barriers, giant resonances, and heavy-ion collisions [2].

In this study, it is investigated the nuclear ground state properties of ^{209}Bi by using *Skyrme-Hartree-Fock* (SHF) method with *Skyrme* set parameters, i.e. *SLy4*, *SkM**, Z_σ, and *SIII*. It is focused to calculated the charge, mass, proton, and neutron radii, the total energy, and the neutron skin thickness. The nuclear charge density is a most useful observable for analyzing nuclear structure: it provides information about the nuclear shape and can be determined by clear-cut procedures from the cross section for elastic electron scattering [3]. The calculation results have been compared with the experiment and theoretical results of the other researchers.

2. Theory

2.1. Skyrme-Hartree-Fock

In 1950, T.H.R Skyrme [4] first proposed the effective interaction between nucleons by a zero-range potential. In this phenomenological interaction, the nucleons potential can be represented by

$$\begin{aligned} V_{Skyrme} &= t_0\left(1+x_0P_x\right)\delta\left(\vec{r}_{ij}\right)+\tfrac{1}{2}t_1\left(1+x_1P_x\right)\left\{\vec{p}_{12}^{\,2}\,\delta(\vec{r}_{ij})+\delta(\vec{r}_{ij})\,\vec{p}_{12}^{\,2}\right\}+t_2\left(1+x_1P_x\right)\vec{p}_{12}\cdot\delta\left(\vec{r}_{ij}\right)\vec{p}_{12} \\ &+\tfrac{1}{6}t_3\left(1+x_3P_x\right)\rho^{\alpha}\left(\vec{r}\right)\delta\left(\vec{r}_{ij}\right)+\mathrm{i}\,t_4\vec{p}_{12}\cdot\delta\left(\vec{r}_{ij}\right)\left(\vec{\sigma}_i+\vec{\sigma}_j\right)\times\vec{p}_{12} \end{aligned} \quad (1)$$

where t_0, t_1, t_2, t_3, x_0, x_1, x_2, x_3, α, dan W_0 are the *Skyrme* set parameters, $\vec{p}_{12} = \vec{p}_i - \vec{p}_j$ is the relative momentum, P_x is the space exchange operator $\vec{r}_i \leftrightarrow \vec{r}_j$, $\bar{r} = \frac{1}{2}(\vec{r}_i + \vec{r}_j)$, ρ represents the density, $\delta(\vec{r})$ is the delta is delta function, and $\vec{\sigma}$ is the vector of Pauli spin matrices [5].

The *Hartree-Fock* equations for *Skyrme* interaction are obtained by writing that the total energy E is stationary with respect to individual variations of the single-particle sates φ_β where the single-particle wave functions φ_β have to satisfy the following set equations

$$\left[-\vec{\nabla}\cdot\frac{\hbar^2}{2m_q^*(\vec{r})}\vec{\nabla} + U_q(\vec{r}) + \vec{W}(\vec{r})\cdot(-i)\left(\vec{\nabla}\times\vec{\sigma}\right)\right]\varphi_\beta = e_\beta\varphi_\beta \quad (2)$$

where q stands for the charge of the single-particle state i [1]. The *Hartree-Fock* method with complete formula can be found in these following references [2, 5, 6].

2.2. *The Nuclear Charge Density*

To compute the observable charge density from the *Hartree-Fock* results, one has to take into account that the nucleons themselves have an intrinsic electromagnetic structure [7]. It is needed to fold the proton and neutron densities from the *Hartree-Fock* method with the intrinsic charge density of the nucleons. Folding becomes a simple product in Fourier space, so that it can be transformed the densities to the so-called form factors

$$F_q(k) = 4\pi\int_0^\infty dr\, r^2 j_0(kr)\rho_q(r) \quad (3)$$

where j_0 is the spherical *Bessel* function of *zeroth* order. The charge form factor is then given by

$$F_C(k) = \sum_q\left[F_q(k)G_{E,q}(k) + F_{ls,q}(k)G_M(k)\right]\times\exp\left(\tfrac{1}{8}(\hbar k)^2/\langle P_{cm}^2\rangle\right) \quad (4)$$

where $F_{ls,q}$ is the form factor of the spin-orbit current ∇J (accounting for magnetic contributions to the charge density), $G_{E,q}$ are the electric form factors of the nucleons, G_M is the magnetic form factor of the nucleons (assumed to be equal for both species), and the exponential factor takes into account an unfolding of the spurious vibrations of the nuclear center of mass in the harmonic approximation (the $\langle P_{cm}^2\rangle$ therein is the same as in the zero-point energy). The charge density is obtained from the charge form factor by the inverse *Fourier-Bessel* transform

$$\rho_C(r) = \frac{1}{2\pi^2}\int dk\, k^2 j_0(kr)F_C(k) \quad (5)$$

Other information can be drawn directly from the form factor. The form factor is thus given on a grid $\{k_j, j = 1, \ldots, N_k\}$. All three form parameters defined in the previous subsection require access to the form factor F_C at any value of k. These intermediate values are computed by *Fourier-Bessel* interpolation

$$F(k) = \frac{\sin(kR_{max})}{kR_{max}}\sum_{j=1}^{N_k} -1^{j+1}\left[1 - \frac{kR_{max}}{j\pi}\right]^{-1} F(k_j) \quad (6)$$

where $R_{max} = N_r\Delta_r$ is the length of the coordinate grid [4]. The root-mean-square (rms) radii of charge, neutron and proton can be defined as [8]

$$r_q = \langle r_q^2\rangle^{\frac{1}{2}} = \left[\frac{\int \vec{r}^2\rho_q(\vec{r})d^3\vec{r}}{\int \rho_q(\vec{r})d^3\vec{r}}\right]^{\frac{1}{2}} \quad (7)$$

The neutron skin thickness can be defined as the difference between the neutron rms radius and the proton rms radius $(t = r_n - r_p)$ [8].

3. Research Method

In this research, the calculations of nuclear total energy of ^{209}Bi have been performed by using *Hartree-Fock* method with *Skyrme* forces set parameters, especially *SLy4*, *SkM**, Z_σ, and *SIII*. Calculations process were performed using HAFOMN code [5]. Procedure of HAFOMN code can be seen in Figure 1. Calculations results of this research have been compared with experiment [8, 9] and Tel *et al* calculation results [8]. In this research, the used *Skyrme* force parameters are *SLy4* [11], *SkM** [12], Z_σ [13], and *SIII* [14]. The values of each set parameter can be found in Table 1.

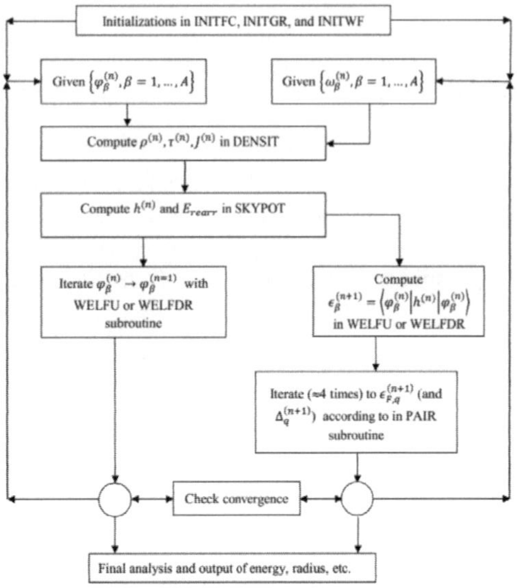

Figure 1. The flowchart of *HAFOMN* code.

Table 1. The *SLy4* [11], *SkM** [12], Z_σ [13], and *SIII* [14] set parameters for *Skyrme* interaction.

Parameter	*SLy4*	*SkM**	Z_σ	*SIII*
t_0	-2488	-2645	-1983.76	-1128.75
t_1	486.82	410	362.252	395
t_2	-546.3	-135	-104.27	-95
t_3	13777	15595	11861.4	14000
x_0	0.834	0.09	1.1717	0.45
x_1	-0.344	0	0	0
x_2	-1	0	0	0
x_3	1.354	0	1.762	1
W_0	123	130	123.69	120
α	0.167	0.167	0.25	1

4. Result and Discussion

4.1. Total energy and Radii Calculation

In this research, it is calculated the total energy for ^{209}Bi by using *Skyrme-Hartree-Fock* method and compared with experiment data taken from *Atomic Data and Nuclear Data Tables* [9]. It can be seen from Table 2 and Table 3 that total energy calculated by Z_σ are in very good agreement with experiment results, followed by *SkM**, *SLy4*, and *SIII* respectively. In this research, after calculating the total energy, it is also investigated the charge, mass, proton, and neutron radii for each nucleus.

Table 2. Total energy calculation results of ^{209}Bi by using *SLy4* and *SkM** set parameters.

Nucleus	Exp [9]	SLy4			SkM*		
		This research	Discrepancy		This research	Discrepancy	
	(MeV)	(MeV)	(MeV)	(MeV)	(MeV)	(MeV)	(MeV)
^{209}Bi	-1640.229	-1622.5549	-17.6741	1.0775	-1629.0986	-11.1304	0.6786

Table 3. Total energy calculation results of ^{209}Bi by using Z_σ and *SIII* set parameters.

Nucleus	Exp [9]	Z_σ			SIII		
		This research	Discrepancy		This research	Discrepancy	
	(MeV)	(MeV)	(MeV)	(MeV)	(MeV)	(MeV)	(MeV)
^{209}Bi	-1640.229	-1641.5865	1.3575	0.0828	-1620.3451	-19.8839	1.2123

From Table 4, all parameters are in good agreement with experiment results, however *SkM** results are the closest to experiment results, followed by the results of *SLy4*, Z_σ, and *SIII* respectively.

Table 4. Calculation results of charge radius of ^{209}Bi (all units are in fm).

Nucleus	This research				Tel et al [8]		Exp [10]
	SLy4	SkM*	Z_σ	SIII	SIII	SkM*	
^{209}Bi	5.539	5.531	5.462	5.597	5.5942	5.5149	5.5211 ± 0.0906

Table 5. Calculation results of mass and proton radii of ^{209}Bi (all units are in fm).

Nucleus	Mass Radius				Proton Radius			
	SLy4	SkM*	Z_σ	SIII	SLy4	SkM*	Z_σ	SIII
^{209}Bi	5.575	5.573	5.439	5.616	5.486	5.478	5.411	5.548

Table 6. Calculation results of neutron radius of ^{209}Bi (all units are in fm).

Nucleus	Neutron Radius				Neutron skin thickness			
	SLy4	SkM*	Z_σ	SIII	SLy4	SkM*	Z_σ	SIII
^{209}Bi	5.634	5.635	5.457	5.661	0.148	0.157	0.046	0.113

4.2. Local Density and Local Potential Distributions

In this research, it is also calculated the distribution of both proton and neutron local densities for each nucleus to understand its ground state properties. The local density distribution for proton and neutron of ^{209}Bi can be seen in Figure 2, where it can be seen that, in radius 0-1 fm, *SLy4* gives the highest results and SIII gives the lowest results. In this area, *SLy4* results decline more drastically than the other parameters. In radius 2-5 fm, Z_σ gives the highest results and SIII gives the lowest results. The results of *SLy4* and *SkM** are similar to each other in this area. In radius 5-12 fm, all parameters give similar results. From neutron density profiles, in radius 0-1 fm, Z_σ obtains the highest results and *SIII* obtains the lowest results. In this area, *SLy4* slope is more drastically than the other parameters. In radius 2-5 fm, Z_σ obtains the highest results and *SIII* obtains the lowest results. The *SLy4* and *SkM** results are more similar in this area. In radius 5-12 fm, all parameters obtain similar pattern but Z_σ is little bit different from the others.

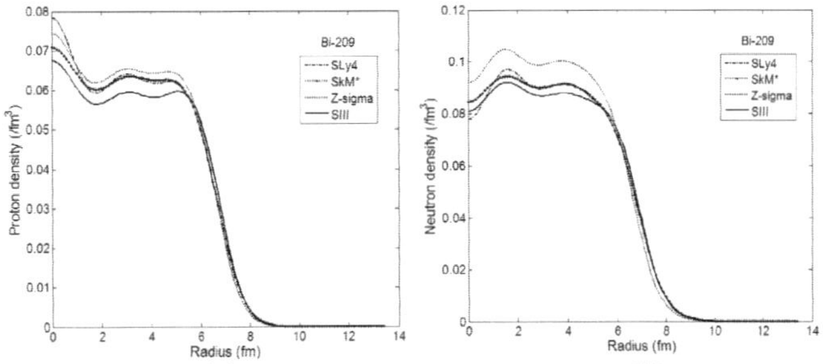

(a) (b)

Figure 2. Nucleon local density distribution of ^{209}Bi obtained by each set parameter.

Figure 3 displays the local potential distributions for proton and neutron of ^{209}Bi. From this figure, it can be seen that *SLy4* obtains the deepest valley. The *SIII* and the *SkM** obtain the same depth of valley but the slopes and surface regions for all parameters are good enough similar to each other. It can be summarized that all parameters obtain similar results, especially the slopes and surface regions (in region >6 fm). All parameter used in this research are in good agreement with the related experiment and the other researchers results. Most of *SkM** results are closer to the related experiment results than *SLy4*, Z_σ, and *SIII* results. From the nucleon density and potential profiles, it can be seen that the obtained results of all parameters are similar to each other, especially the slope and the surface regions. From those results above, it can be indicated that *Skyrme-Hartree-Fock* method can be the useful tool to study the nuclear ground state properties of heavy nuclei.

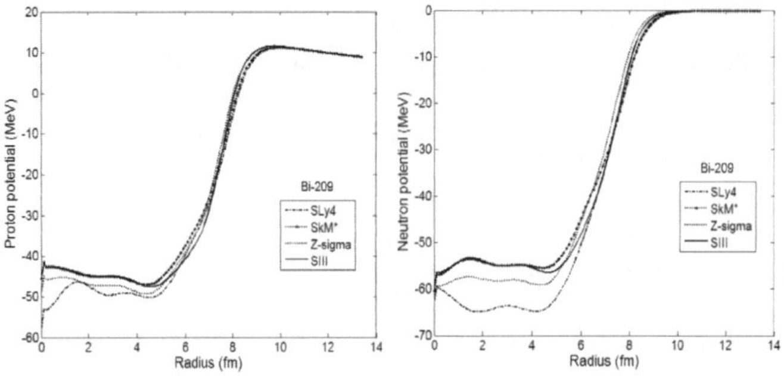

(a) (b)

Figure 3. Nucleon local potential distribution of ^{209}Bi obtained by each set parameter.

5. Conclussion

It has been performed a study to calculate the densities, the neutron skin thickness, and the rms radii for both proton and neutron of ^{209}Bi by using *Hartree-Fock* method with *Skyrme* set parameters, especially *SLy4*, *SkM**, Z_σ, and *SIII* set parameters. In this study, all used parameters are in good agreement with the results of the related experiment and the other researchers, where *SkM** results are closer collectively to experiment energy than *SLy4*, Z_σ, or *SIII*. The radii calculation of this research with *SkM** and *SIII* set parameters are close enough to Tel *et al* and the related experiment results. These results indicate that *Skyrme-Hartree-Fock* method can be the good useful tool to explain the ground state properties of heavy nuclei.

Acknowledgment

The authors are grateful to Prof. P.G. Reinhard (Germany) for discussion and guide in using the *HAFOMN* code.

References

[1] Reinhard P G and Nazarewicz W *arXiv:1601.06324v1*
[2] Vautherin D and Brink D M 1972 *Phys. Rev. C* **5** 626
[3] Dreher B, Friedrich J, Merle K, Luhrs G and Rothaas H 1974 *Nucl. Phys. A* **235** 219
[4] Skyrme T H R 1959 *Nucl. Phys.* **9** 615
[5] Reinhard P G 1991 *Skyrme-Hartree-Fock Model of the Nuclear Ground State* (Computational Nuclear Physics I - Nuclear Structure) ed. Langanke K, Koonin, S E and Maruhn J A (Springer-Verlag Berlin Heidelberg) chapter 2 pp 28-50
[6] Bender M, Heenen P H, and Reinhard P G 2003 *Reviews of Modern Physics* **75** 121
[7] Friar J L and Negele J W 1975 *Adv. Nucl. Phys.* **8** 219
[8] Tel E and Aydin A 2012 *J. Fusion Energy* **31** 73
[9] Wang M *et al.* 2012 *Chinese Physics C* **36** 1603
[10] Angeli I and Marinova K 2013 *Atomic Data and Nuclear Data Tables* **99** 69
[11] Chabanat E, *et al.* 1997 *Nucl. Phys. A* **627** 710
[12] Bartel J, Quentin P, and Brack M 1982 *Nucl. Phys. A* **386** 79
[13] Friedrich J and Reinhard P G 1986 *Phys. Rev. C* **33** 335
[14] Beiner M, Flocard H, and Giai N V 1975 *Nucl. Phys. A* **238** 29

IOP Conf. Series: Journal of Physics: Conf. Series **799** (2017) 012025 doi:10.1088/1742-6596/799/1/012025

Design Study of Modular Nuclear Power Plant with Small Long Life Gas Cooled Fast Reactors Utilizing MOX Fuel

Muhammad Ilham[1]**, Zaki Su'ud**
Nuclear Physics and Biophysics Research Division, Physics Department, Faculty of Mathematics and Natural Science, Bandung Institute of Technology Indonesia, Jalan Ganesha 10 Bandung 40132, INDONESIA

E-mail: muhilham99@gmail.com

Abstract. Growing energy needed due to increasing of the world's population encourages development of technology and science of nuclear power plant in its safety and security. In this research, it will be explained about design study of modular fast reactor with helium gas cooling (GCFR) small long life reactor, which can be operated over 20 years. It had been conducted about neutronic design GCFR with Mixed Oxide (UO2-PuO2) fuel in range of 100-200 MWth NPPs of power and 50-60% of fuel fraction variation with cylindrical pin cell and cylindrical balance of reactor core geometry. Calculation method used SRAC-CITATION code. The obtained results are the effective multiplication factor and density value of core reactor power (with geometry optimalization) to obtain optimum design core reactor power, whereas the obtained of optimum core reactor power is 200 MWth with 55% of fuel fraction and 9-13% of percentages.

Introduction

Global demand for energy continues to increase, the world's population increases more rapidly and the sources of non-renewable energy tend to be decreased are important factors in effort to develop a new generation of power plant with advanced technology that can generate considerable amount of energy. The Rapid development in technology and science in the field of energy create an alternative energy source by nuclear reactors which utilize a controlled nuclear reaction. Indonesia have a serious problem with electricity shortage in some areas mainly outside Java and Bali islands. These problems show a possibility to introduce nuclear power plants in order to increase the electrical production. Nuclear power plants with long life characteristic are very prospective for remote areas, In such area small reactors are best fit [8]. But this alternative energy source has a lot of resistance in the community of the world, including in Indonesia. The accidents of Chernobyl and Fukushima nuclear while ago cause negative opinion and public rejection in the construction of nuclear power plants [6].
The development of nuclear power plants has reached a phase of fourth generation and certainly much better than the previous generation. The fourth generation nuclear power plants are alternative energy sources with high density of energy, better conversion energy to electricity and good for environment because it does not contribute to global warming gaseous production. These advanced NPPs can also resolve many problems from the early generation NPPs by have much higher efficiency using natural uranium as a fuel, passive safety, and lower energy prices [2-5].

Gas cooled fast reactors (GCFR) are among fourth generation NPPs with fast neutron spectrum and have excellent capability to utilize natural uranium during operation [1]. In this research study of small long life gas cooled fast reactors with Mixed Oxide fuel (MOX), helium as coolant, with total power level up to 200 MWth will be simulated. This nuclear power plants (NPPs) that has cylindrical pin cell and cylindrical balance of reactor core geometry will be optimized to obtain reactors that can be operate in long period with low reactivity swing.

Experimental Method

In general, the placement of the reactors under the surface can be realized with two different basic construction concepts, i.e. placing the entire section of the reactor and turbine in deep rock containment with below normal level in relation to water recipient and placing the reactor in deep rock containment with turbine in above ground building at normal level in relation to water recipient.

F1	F1	F2	F2	F3	F3	R
F1	F1	F2	F2	F3	F3	R
F1	F1	F2	F2	F3	F3	R
F1	F1	F2	F2	F3	F3	R
F1	F1	F2	F2	F3	F3	R
F1	F1	F2	F2	F3	F3	R
R	R	R	R	R	R	R

Figure 1 Configuration of reactor core

The reactor core was subdivided into four regions in radial direction with cylindrical pin cell and cylindrical balance of reactor core geometry. In Figure 1, three regions belong to fuel with different percentages of plutonium (F1, F2, F3) and one to reflector (R) with equal length in each fuel region. The outer region core has more plutonium percentages than inner core to obtain power density to be spread evenly across region of reactor core.

This paper will be focused on the parameters design of reactors core which is neutronic calculation. The calculation neutronic analysis were performed using SRAC code program (SRAC-CITATION code system) [7] with nuclides data by library of JENDL-3.2. In the SRAC code, the burn up calculations were performed using cell burn-up module to obtained the data energy group macroscopic cross section to be used in 2D R-Z geometry multi-groups diffusion calculation using CITATION module. The diffusion calculation will produce the data average power density of each region that will be reused into SRAC code for cell burn-up calculation, this process will be repeated until the result of data power density is converged

Result and Discussion

In table 1 shows design parameters of reactor with three variation were calculated at 100 MWth, 150 MWth and 200 MWth power level and fuel volume fraction were 50%, 55%, 60% using plutonium percentages at 9%, 11%, 13% of MOX fuel in region F1, F2, and F3.

Table 1. General parameters of the design core reactors

Parameter	Value
Power (MWth)	100-200
Core geometry	Cylindrical balance
Number of radial division of the main core	4 regions
Operation time without refueling (years)	20
Fuel / Cladding / Coolant type	MOX / SS316 / He gas
Active core radius / height (cm)	216 / 216
Reflector type / width (cm)	Stainless steel / 50
Plutonium percentages (%)	9-13
Fuel volume fraction (%)	50-60
Cladding volume fraction (%)	10
Coolant volume fraction (%)	10-20

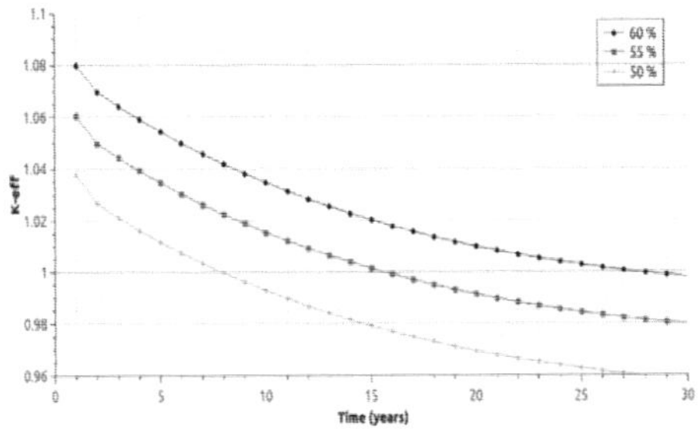

Figure 2. Figure 2 Effective multiplication factor changes over time in case of 100 MWth power level with variation in fuel fraction.

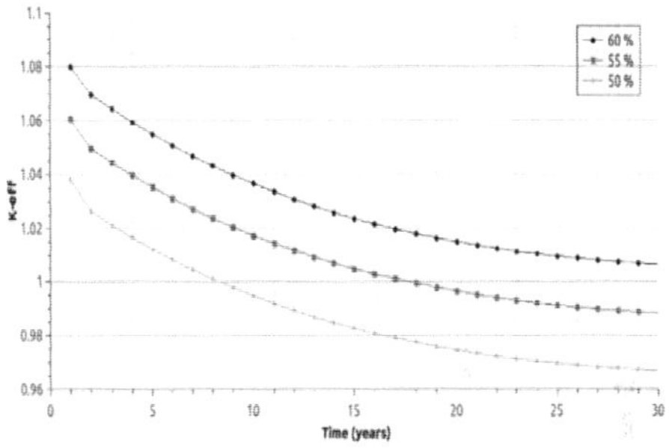

Figure 3. Effective multiplication factor changes over time in case 150 MWth power level with variation in fuel fraction.

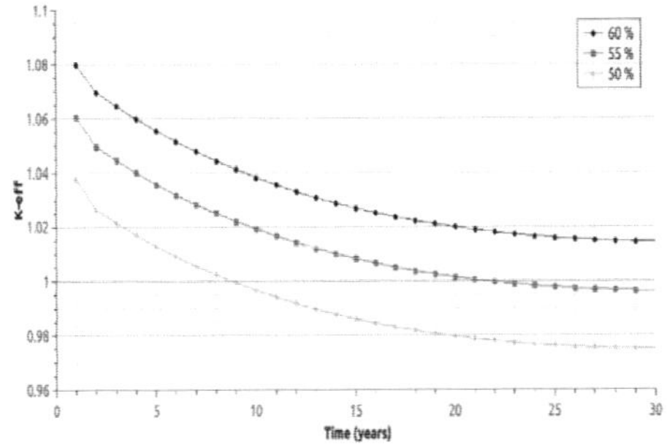

Figure 4. Effective multiplication factor changes over time in case of 200 MWth power level with variation in fuel fraction.

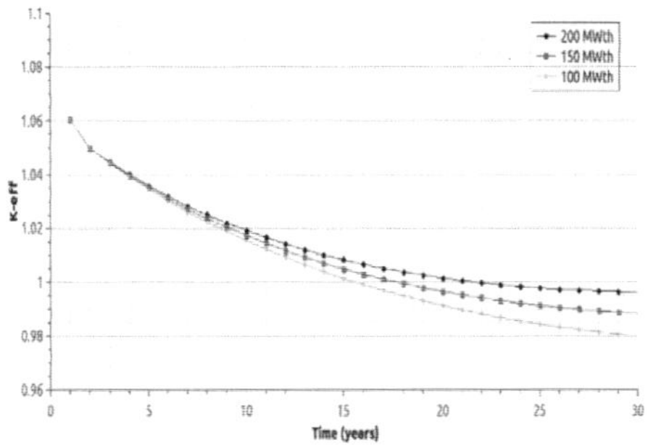

Figure 5. Effective multiplication factor changes over time in case of 55% MWth fuel fraction with variation in power level.

Figure 3 and 4 show the change in value of effective multiplication factor (K-eff) during burn-up in case of 100 MWth and 150 MWth reactor power level with variation in fuel volume fraction by 50 % , 55 % and 60 % using percentages of fuel plutonium by 9 % , 11 % and 13 % of the total fuel to the region F1 , F2 , F3. The results of 100 MWth's case shown that the value of effective multiplication factor steadily decreasing from beginning of life (BOL) to the end of life (EOL). By decreasing and increasing the fuel fraction parameters by 5%, the result will have a similar pattern. The effect for decreasing and increasing fuel fraction will shift down and shift up the K-eff pattern. The higher fuel fraction makes the longer operation of reactor. In case of 150 MWth's case, it has shown the effect for decreasing and increasing fuel fraction also shift down and shift up the K-eff pattern, same as the previous case but it have a few year longer.

Figure 5 show the change in value of effective multiplication factor (K-eff) during burn-up in case of 200 MWth reactor power level with variation in fuel volume fraction. The variation is also performed by decreasing and increasing the fuel fraction parameters by 5%. It shown that K-eff is still has the same pattern as two cases before. Increasing the fuel fraction makes the period of operating reactor is also increase by few years. In this case the reactor has the longest period of operating that up to over 20 years.

Figure 6 shows the change in value of effective multiplication factor (K-eff) during burn-up in case of 55% fuel fraction with variation in reactor power level by 100, 150 and 200 MWth. The variation is performed by decreasing and increasing the power level parameters by 50 MWth. It shown that the K-eff of three variation power level have similar value and shape at the beginning of life (BOL), has slight change of shape in medium of life (MOL) and major change of shape at end of life (EOL) indicate that the effect of power level make period operating of reactor increased.

Conclusion

Design study of modular nuclear power plant with small long life gas cooled fast reactors utilizing MOX fuel has been performed. The reactors are small size with helium gas cooled, fast neutron spectrum and have long lifetime that could be operate for 20 years without refueling. In this research the modular power plant have result in neutronic design for 100-200 MWth power level and 50-60% fuel fraction has been discussed. To obtain optimum design core reactor, optimum reactor power level is 200 MWth with 55% fuel fraction and 9%, 11%, 13% percentages of plutonium in region F1, F2, F3.

Acknowledgent

This research partially support by advanced nuclear laboratory of ITB.

References

[1] Su'ud, Z., Sekimoto, H., Annals of Nuclear Energy, Vol.54 (2013), pp.58-66
[2] Zaki Su'ud, Design study of small gas cooled fast nuclear power plant for synergetic energy system with renewable energy by employing pump storage, Advanced Material Research (AMR), Vol. 983 (2014) pp 233-237
[3] Z. Su'ud, Progress in Nuclear Energy 32 (3-4) (1998), pp. 571-577
[4] H. Sekimoto and Zaki Su'ud, Nuclear Technology, Vol. 105, no.3 (1995), pp.307-313
[5] Z. Su'ud, Progress of Nuclear Energy, Vol. 50 (2008), pp. 276-278
[6] Zaki Su'ud, The Role of Nuclear Energy in Indonesia in The 21 st Century, GENEs4/ANP 2003, Sep. 15-19, 2003, Kyoto, Japan, pp 1227
[7] K. Okumura et al. SRAC (Ver.2002); The comprehensive neutronics calculation code system, JAERI report (2002)
[8] M. Ariani, Z. Su'ud, F. Monado, A. Waris, Khairurrijal , I. Arif , A. Ferhat , and H. Sekimoto, Optimization of Small Long Life Gas Cooled Fast Reactors With Natural Uranium as Fuel Cycle Input, Applied Mechanics and Material (AMM) , Vol. 260-261 (2013) pp 307-311

IOP Conf. Series: Journal of Physics: Conf. Series **799** (2017) 012026 doi:10.1088/1742-6596/799/1/012026

Analysis on Reactor Criticality Condition and Fuel Conversion Capability Based on Different Loaded Plutonium Composition in FBR Core

Sidik Permana[1*], Geby Saputra[1], Mitsutoshi Suzuki[2], Masaki Saito[3]

[1] Nuclear Physics and Bio Physics Research Group, Bandung Institute of Technology, Gedung Fisika, Jl. Ganesha 10, Bandung 40132, Indonesia.
[2] Department of Science and Technology for Nuclear Material Management (STNM), Japan Atomic Energy Agency (JAEA), 2-4 Shirane, Shirakata, Tokai Mura, Naka- gun, Ibaraki 319-1195 Japan.
[3] Research Laboratory for Nuclear Reactors, Tokyo Institute of Technology 2-12-1 O-Okayama, Meguro-ku, Tokyo 152-8550, Japan

Email: psidik@fi.itb.ac.id

Abstract. Reactor criticality condition and fuel conversion capability are depending on the fuel arrangement schemes, reactor core geometry and fuel burnup process as well as the effect of different fuel cycle and fuel composition. Criticality condition of reactor core and breeding ratio capability have been investigated in this present study based on fast breeder reactor (FBR) type for different loaded fuel compositions of plutonium in the fuel core regions. Loaded fuel of Plutonium compositions are based on spent nuclear fuel (SNF) of light water reactor (LWR) for different fuel burnup process and cooling time conditions of the reactors. Obtained results show that different initial fuels of plutonium gives a significant chance in criticality conditions and fuel conversion capability. Loaded plutonium based on higher burnup process gives a reduction value of criticality condition or less excess reactivity. It also obtains more fuel breeding ratio capability or more breeding gain. Some loaded plutonium based on longer cooling time of LWR gives less excess reactivity and in the same time, it gives higher breeding ratio capability of the reactors. More composition of even mass plutonium isotopes gives more absorption neutron which affects to decresing criticality or less excess reactivity in the core. Similar condition that more absorption neutron by fertile material or even mass plutonium will produce more fissile material or odd mass plutonium isotopes to increase the breeding gain of the reactor.

1. Introduction

Recycling option of spent nuclear fuels option from discharged LWR fuels is one of the important issues to be pursued because of the potential to extend the sustainability of nuclear fuel such as recycling program of used uranium and plutonium, instead of once through fuel option. Other spent nuclear fuel material such as minor actinides or MAs can be utilized also for recycling nuclear fuel program in order to reduce spent nuclear fuel (SNF) composition and to increase the level of nuclear proliferation resistance. Fast reactor technology which uses fast neutron spectrum region, shows better fuel breeding capability and at the same time, it has a better level of reducing spent nuclear fuels by burning process. It also show better criticality condition as well as higher fuel breeding because of high eta value of fissile materials at fast neutron energy region. Minor actinide (MA) material as transuranium material is also can be used to produce protected plutonium proliferation which can improve nuclear nonproliferation level [1-5]. Several recycling programs of SNF have been extensively conducted for long life core reactor program [6-7], increasing burning MA rate [8], and better transmutation TRU rate [9] and a program for recycling MA in light water reactor (LWR) type [10]. Reactor criticality condition and nuclear fuel breeding ratio are evaluated in the present study which is based on fast breeder reactor (FBR) type. Fuel loading are based on some different supply fuels which is mainly from different loaded plutonium composition in the fuel core regions.

2. Methods

Analysis on LWR fuel composition have been conducted by using depletion code of ORIGEN computer code [11] for different fuel burnup condition and cooling time process. Those LWR SNF compositions are used for initial loaded nuclear fuel of FBR. In term of FBR evaluations such as criticality condition, fuel composition and conversion ratio capability of the reactor have performed by using a coupling codes including SLAROM code, JOINT and CITATION codes. In this evaluation, nuclear data library was using JFS-3-J-3.2R for nuclear data analysis of investigated reactor systems which is based on the JENDL 3.2 [12-15]. JSFR (Japan Sodium Fast Reactor) design was used as the basis design for this evaluation [16].

2.1. Breeding Ratio

In this investigation, a parameter of nuclear fuel sustainability will be based on nuclear fuel breeding rasio which is based on some reaction rates of some fissile and fertile materials of nuclear fuels as shown in **Eq. 1.**

$$BR = \frac{\left(\sum_{i=1}^{i=n} \Sigma_{c,i}^{fertile}\right)}{\left(\sum_{i=1}^{i=n} \Sigma_{a,i}^{fissile}\right)_v} \qquad (1)$$

$\Sigma_{c,i}^{fertile}$: Macroscopic capture cross-section of i isotope for fertile materials

$\Sigma_{a,i}^{fissile}$: Macroscopic absorption cross-section of i isotope for fissile materials

As mention in Eq. 1, ratio of breeding is based on a ratio between capture cross-section of fertile materials and absorption cross-section of fissile materials. Main fertile materiasl are U-238, Pu-240 and Pu-238 and fissile materials are U-235, Pu-239 and Pu-241. Some different plutonium compositions in the core as initial loaded fuel will give a signify contribution to reactor performances especially for breeding ratio capability and criticality condition during reactor operation.

3. Results and Discussion

Obtained results of uranium and plutonium based on SNF of LWR are shown in Table 1 and 2 for different fuel burnup process and cooling time. Isotopic vector composition has some differences depending on burnup and cooling time. Some plutonium isotopes are increasing and some others are reducing for longer cooling time due to the half-life of the plutonium isotopes and conversion process from nuclide to another nuclide. Higher burnup affects to increases all isotopic plutonium composition except for Pu-239 which is reducing for higher burnup. Reducing Pu-239 is estimated from fission reaction from Pu-239 for reactor operation.

Table 1 Plutonium composition of spent nuclear fuel composition LWR for different cooling time.

Pu Vector Comp. of LWR 50 GWd/t		
Plutonium	Cooling Time [Years]	
	5.0	30.0
Pu-238	3.00	2.70
Pu-239	52.62	57.63
Pu-240	24.50	27.23
Pu-241	12.20	4.01
Pu-242	7.69	8.42

Table 2 Plutonium composition of spent nuclear fuel composition LWR for different burnups.

Pu Vector Comp. of LWR 33-50 GWd/t		
Plutonium	Cooling Time 30 [Years]	
	33	50.0
Pu-238	1.464	2.704
Pu-239	63.618	57.632
Pu-240	25.757	27.227
Pu-241	3.557	4.014
Pu-242	5.605	8.423

3.1 Criticality Operation Condition

Reactor operation is directly related to a criticality conditions of the reactor which is based on the fission reaction that occurs during reactor operation. Obtained some criticality conditions are shown in **Figures 1** and **2** for different fuel loading in the core reagion. As mentioned before it is estimated some different fuel composition will affect to the reactor criticality condition especially from the initial fuel loading. It shows some different criticality conditions of FBR during reactor operation up to equilibrium reactor condition for different composition of loading fuel based on different cooling time process and fuel burnup. Initial criticality condition is estimated higher than equilibrium fuel composition because of more fission occur for initial fuel condition and it becomes less when fissionable materials are decreasing as well as production of some nuclide which absorp neutron. In term of different cooling time, it shows that longer cooling time of loaded SNF LWR in the FBR gives a reduction value of criticality condition at the beginning of cycle (BOC) and equilibrium condition. In initial reactor condition, less criticality means less reactor reactivity that stability of neutron population can be reduced.

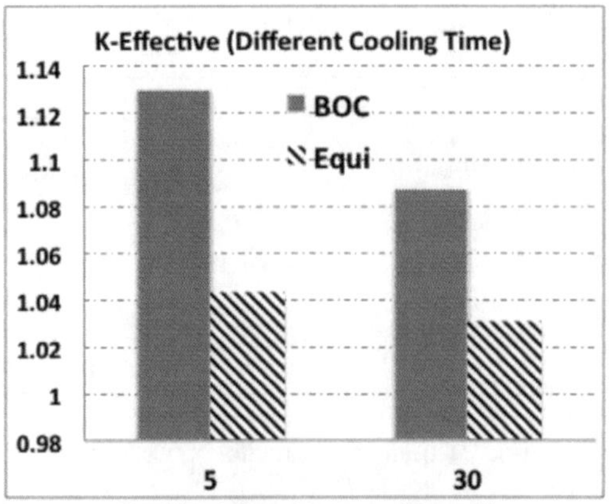

Cooling Time Process [Years]

Figure 1. Criticality condition of FBR core based on different cooling time of loaded SNF LWR
BOC: Beginning of cycle
Equi: Equilibrium cycle

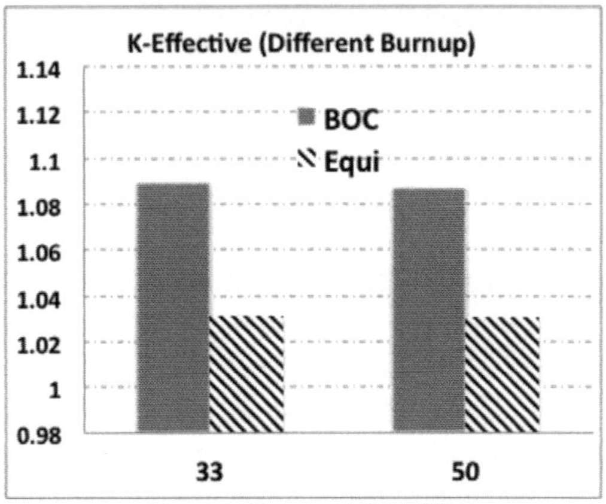

Burnup Process [GWd/t]

Figure 2. Criticality condition based on different burnup process of loaded SNF

It is expected to have less criticality condition for higher burnup as shown in Figure 2 that more burnup value obtains less criticality condition for both BOC and equilibrium condition. Longer burnup is estimated to have less fissile material such as fissile plutonium for maintaining reactor operation. Based on the results, more cooling time process is used and higher level of fuel burnup is increased will obtain a better safety condition in term of less excess reactivity or less excess neutron production during reactor operation, especially in initial reactor operation.

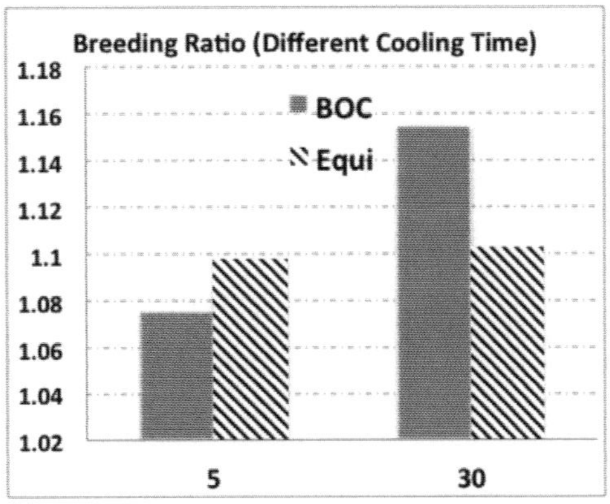

Cooling Time Process [Years]

Figure 3 Conversion Ratio based on different cooling time of loaded LWR

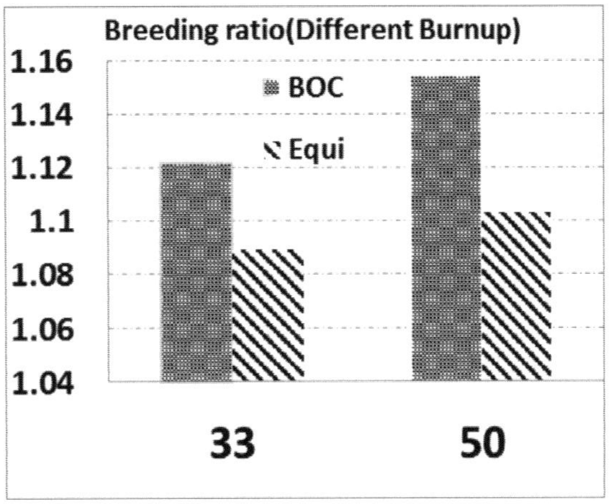

Burnup Process [GWd/t]

Figure 4. Conversion Ratio condition based on different burnup process of loaded SNF LWR.

3.2. Fuel Breeding Ratio

Fuel breeding ratio is evaluated during reactor operation up to equilibrium condition which are shown in Figures 3 and 4 as fuel breeding profile based on different loaded fuel types of SNF LWR into the FBR cycle. Longer reactor operation affect to increases fuel breeding ratio value for both cooling time processes and it shows more significant increasing fuel breeding value for for initial breeding ratio of FBR than at equilibrium condition. Fuel breeding ratio of the reactor is increasing with increasing cooling time and it shows a slightly increase for equilibrium reactor condition._Some loaded actinide SNF composition based on higher burnup obtain better breeding ratio of FBR cycle for all reactor operation such as beginning of operation up to equilibrium condition as shown in Figure 4. Better level of fuel breeding ratio is estimated from the increase of fuel conversion capability of fertile materials into fissile materials. In relation to criticality condition, less criticality obtains better fuel breeding. This condition can be expected from capturing more neutron will reduce excess reactivity which corresponding to reactor criticality. At the same time, capturing neutron by actinide was used to convert some fertile materials into fissile materials during reactor operation.

4. Conclusions

Reactor criticality condition and nuclear fuel breeding ratio have been investigated based on fast breeder reactor (FBR) type. Initial fuel loading are based on some different supply fuels of spent nuclear fuel of LWR which is mainly from different loaded plutonium composition in the fuel core regions. Some plutonium isotopes are increasing and some others are reducing for longer cooling time due to the half-life of the plutonium isotopes and conversion process from nuclide to another nuclide. Higher burnup affects to increases all isotopic plutonium composition except for Pu-239 which is reducing for higher burnup. Initial criticality condition is estimated higher than equilibrium fuel composition. Longer cooling time process and higher level of fuel burnup give a better safety condition in term of less excess reactivity or less excess neutron production during reactor operation, especially in initial reactor operation.

Longer reactor operation affect to increases fuel breeding ratio and it increases with increasing cooling time. Loaded actinide composition based on higher burnup of SNF LWR obtain better breeding ratio of FBR cycle for all reactor operation. Higher fuel breeding ratio is estimated from the increase of fuel conversion capability of fertile materials into fissile materials. In relation to criticality condition, less criticality obtains better fuel breeding. This condition can be expected from capturing more neutron will reduce excess reactivity which corresponding to reactor criticality. At the same time, capturing neutron by actinide was used to convert some fertile materials into fissile materials during reactor operation.

Acknowledgment

We would like acknowledge and extend our gratitude to ITB research innovation program and Desentralisasi research program of ministry of research, technology and higher education for the international conference grant and publication.

References

[1] M Saito 2002 *Prog. in Nuc. Energy* 40 (3-4) 365-374
[2] International Atomic Energy Agency 1972 Information Circular, INFCIRC/153
[3] B. Pellaud 2002 *J. Nucl. Mater. Managements* XXXI(1) 30-38
[4] G Kessler 2007 *Nucl. Sci. and Eng.* 155 53-73
[5] P. Sidik 2011 *Nucl. Eng. and Des.* 241 101-117
[6] Sidik Permana Zaki Su'ud 2009 Core Performance and Plutonium Production of Small Long Life Fast Reactor using Doping Actinides, *Proceeding of International Congress on Advance In Nuclear Power Plants (ICAPP)* Tokyo-Japan
[7] Sidik Permana Zaki Suud and Mitsutoshi Suzuki 2009 Trans-uranium doping utilization for increasing protected plutonium proliferation of small long life reactor *Proceedings of Global 2009 "The Nuclear Fuel Cycle: Sustainable Options & Industrial Perspectives"* Paris France
[8] H Choi and T J Downar 1999 A Liquid-Metal Reactor for Burning Minor Actinides of Spent Light Water Reactor Fuel—I: *Neutronics Design Study Nuclear Engineering and Design* 133 1–22 1999
[9] Edward A Hoffman and Weston M Stacey 2003 *Comparative Fuel Cycle Analysis of Critical and Subcritical Fast Reactor Transmutation Systems Nuclear Technology* 144 83
[10] T A Taiwo et al. 2006 *Assessment of a Heterogeneous PWR Assembly for Plutonium and Minor Actinide Recycle Nuclear Technology* 155 34
[11] Oak Ridge Isotope GENeration Code 1980 *"ORIGEN-2, Isotope Generation and Depletion Code"* ORNL TM-7175
[12] M Nakagawa K Tsuchihashi 1984 "SLAROM: A Code for Cell Homogenizations Calculation of Fast Reactor JAERI 1294" *Japan Atomic Energy Research Institute*

[13] T B Fowler D R Vondy G W Cunningham 1971 *"Nuclear Reactor Core Analysis Code: CITATION, Oak Ridge National Laboratory report"* ORNL-TM-2496 Rev2 USA

[14] G Chiba et al. 2002 *J. At. Energy Soc. Jpn.* 14 335

[15] T Nakagawa *et al.* 1995 *J. Nucl. Sci. Technol.* 32 1259

[16] S Ohki et al. 2008 "Design Study of Minor Actinide Bearing Oxide Fuel Core for Homogeneous TRU Recycling Fast Reactor System" *The 10th OECD/NEA P&T Meeting* Mito Japan

Study of gamma spectrometry laboratory measurement in various sediment and vulcanic rocks

Bagus Endar B. Nurhandoko[1,2], Rizal Kurniadi[1,2], Muhammad Rizka Asmara Hadi[2], Insan Rizal Komara[2]

[1]Physics Department, Faculty of Mathematics and Natural Science,Institut Teknologi Bandung
[2]Rock Fluid Imaging Lab.

Email: bagusnur@bdg.centrin.net.id

Abstract Gamma-ray spectroscopy is the quantitative study of the energy spectra of gamma-ray sources. This method is powerful to characterize some minerals, especially to differentiate rocks which contains among Potassium, Uranium, dan Thorium. Rock contains radioactive material which produce gamma rays in various energies and intensities. When these emissions are detected and analyzed with a spectroscopy system, a gamma-ray energy spectrum can be used as indicator for mineral content of rock. Some sediment and vulcanic rock have been collected from East Java Basin. Samples are ranging from Andesite vulcanics, Tuff, Shale, various vulcanic clay and Alluvial clay. We present some unique characteristics of gamma spectrometry in various sedimentar and vulcanic rocks of East Java Basins. Details contents of gamma ray spectra give enrichments to characterize sample of sediment and vulcanic in East Java. Weathered vulcanic clay has lower counting rate of gamma ray than alluvial deltaic clay counting rate. Therefore, gamma spectrometrometry can be used as tool for characterizing the enviroment of clay whether vulcanic or alluvial-deltaic. This phenomena indicates that gamma ray spectrometry can be as tool for characterizing the clay whether it tends to Smectite or Illite

1. Introduction

The database information of gamma radiation of rocks is needed for many purposes such example as radiation measurement and dose rate of tiling rocks[1], identification of low-grade uranium (U) ore with thorium resources in a large syngenetic deposit using gamma spectrometry borehole logging [2], detection of potassic alteration to mineralization [3] and challenge discrimination of clay among Illite, Smectite and Kaolinite.

Some authors measured gamma spectrometry to characterize radioactivity of rock and soils [4,5,6]. In this work, sample rocks collected in East Java Basin are measured to initiate the databases of gamma spectrometry in various sediment rock in Indonesia.

Gamma spectrometer was used widely to identify radionuclide gamma emitter especially to discriminate among Thorium, Uranium and Potassium[7]. As well as Si or Li detectors[8], the scintillation detector NaI (T1) is detector that compose with material that could makes flash light if interact with gamma radiation. In detector, gamma ray makes photoelectric effect, Compton scattering, and pair production. Light intensity as output scintilator crystal proportion with gamma ray energy. Flash light is transmitted to photocathode from photomultiplier tube (PMT). Electric current from

PMT makes pulse in preamplifier. The pulse then is processed with Multichannel Analyzer Analyzer (MCA). Pulse amplitude is proportion with gamma energy.

There are three radionuclide series, Uranium, Thorium, and Potassium series. All kinds of rocks have uranium, especially granite where one ton granite has about three until four gram Uranium. Uranium generally has homogeny distribution in rock. Mainly uranite has 80% and euksinit has 20%. There are uranium isotop with abundances 99,275% (238U), 0,72% (235U), and 0,005% (234U) [8]

There is small amount of Thorium in rocks and soil. In soil Thorium amount about 12 ppm with 232Th isotope has abundance 100%.

Carbonate mainly is composed by Potassium with isotopes are 39K (93,3%), 40K (0,012%) and 41K (6,7%). Isotope 40K is radioactive which it decays into 40Ca and 40Ar with it radiates gamma and beta [9].

Radionuclide detection is easility detected by gamma spectrometer, the uniqueness of energy spectrum can discriminate among Uranium, Thorium, and Pottasium concentration[9][10].

In this work, gamma ray spectrometer is used to detect the existence of Potassium, Uranium, and Thorium from sample of rocks in East Java Basin

2. Experimental Method

The samples of sediment rocks were collected from northern part to southern part of East Java Figure 1.

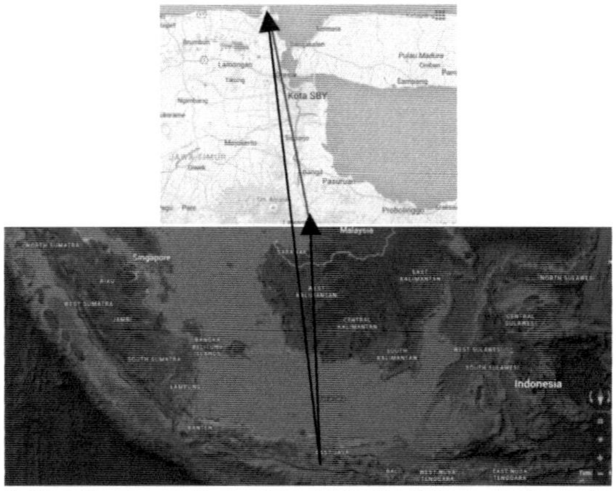

Figure 1 The line sampling of rocks in East Java, Indonesia.

Gamma spectrum from rocks is measured using NaI Scintillator Probe at range energy between 25 KeV – 3500 KeV. This detector is furnished by positive polarity 750 volt pre-amplifier. To reduce background radiation, Pb plate is placed around the device. Block diagram of gamma spectrometry is shown figure 2.

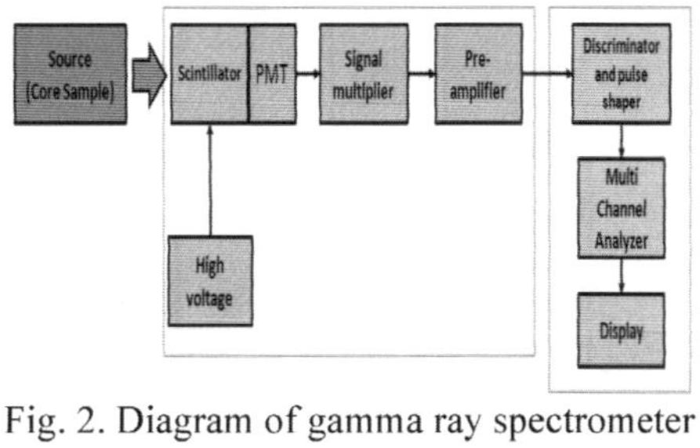

Fig. 2. Diagram of gamma ray spectrometer

IOP Conf. Series: Journal of Physics: Conf. Series **799** (2017) 012027 doi:10.1088/1742-6596/799/1/012027

3. Results and Dicussion

Counting time is choosen 600 second. Figure 3 and 4 shows the counting curve vs energy (in keV) of ten sample of rocks. Type of sample covers: vulcanic claystone, vulcanic tuff, weathered claystone and alluvial clay. The vulcanic claystone, culcanic sand, weathered claystone were collected from Cangar, near Malang. Some samples of claystone where collected near Jolotundo in leg of Mt. Pananggungan. Some samples of vulcanic tuff and vulcanic "alluvial" were collected from Songgoriti and leg of Mt.Welirang, The detail of sample position is listed in Table 1.

Table 1 shows tabulation of total countingrate. The sample of rocks is sorted by countingrate value from higher to lower. The result of data processing from ten sample also is showen by figure 3. In sand rock, counting rate of Potassium is about 140 -192, Uranium about 64 – 87, Thorium 56 – 87. Clay rock samples counting rate are Potassium about 339 – 496, Uranium about 117 – 125, Thorium 191 – 306, and last, Tuff samples and Aluvial samples counting rate are Potassium about 596, Uranium 117, Thorium 360 for Tuff, and Potassium 170, and Uranium about 80 for Aluvial.

The counting data of weathered (soft) clay rock presents as following: the counting rate in the range of Potassium is about 79-161, one of Uranium is about 58 – 66, and one of Thorium about 13 – 19.

Figure 4. shows counting rate of gamma ray in the range of Potassium, Thorium and Uranium from several collected rock samples. The counting rate of samples are dominant in the range of Potassium, range of Uranium is minimum. This phenomena shows uniquenes gamma spectrum of sediment as well as vulcanic rocks in East Java Basin.

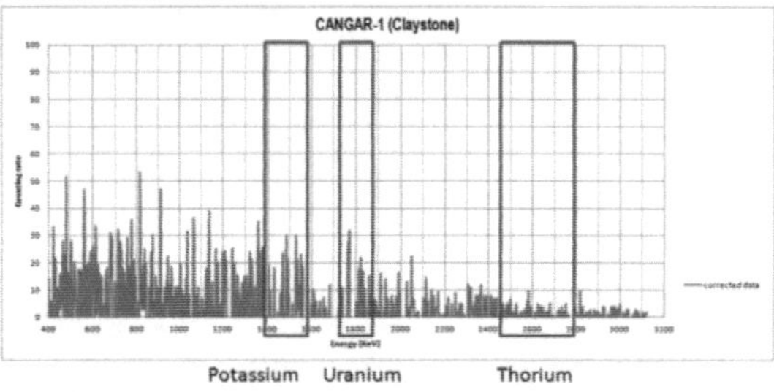

(a)

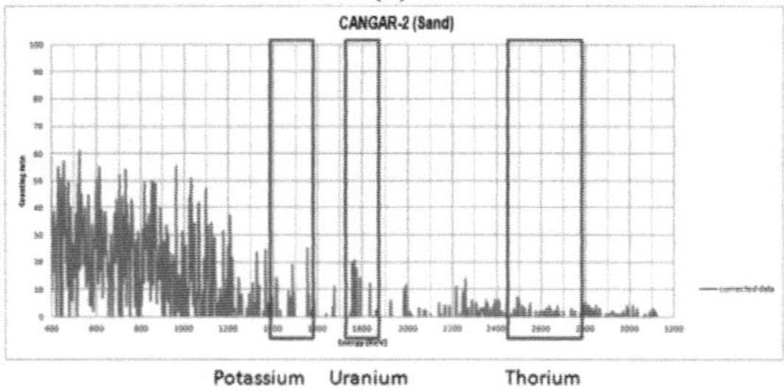

(b)

Fig. 2. Counting vs enery (in Kev) curve from (a) Claystone Cangar-1 (b) Sand Cangar-2.

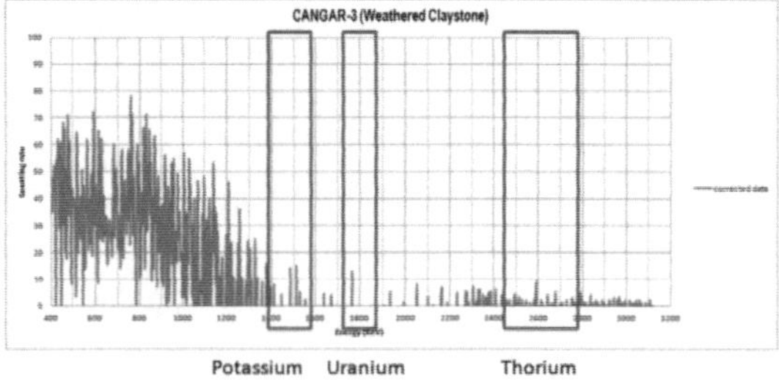

(a)

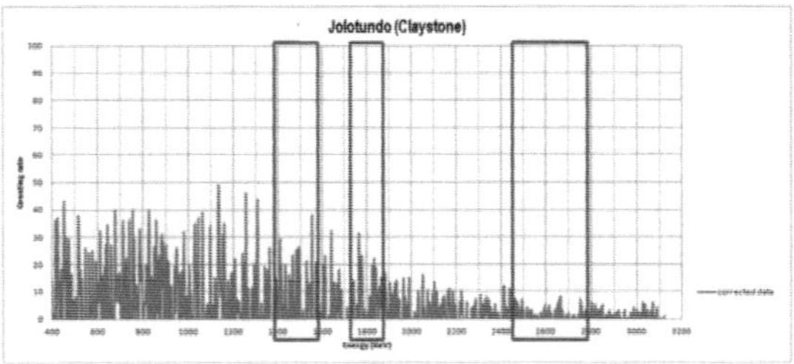

Potassium Uranium Thorium

(b)

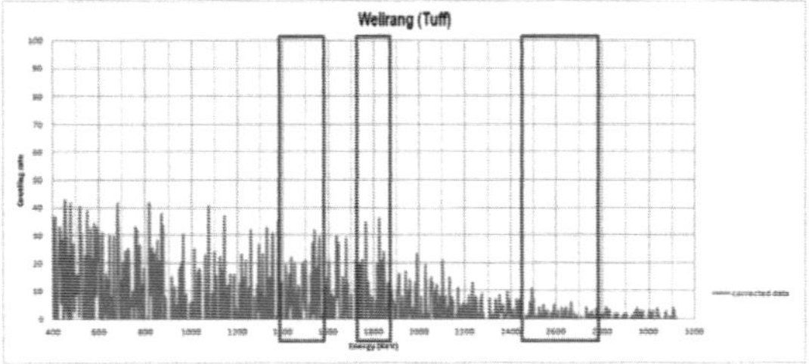

Potassium Uranium Thorium

(c)

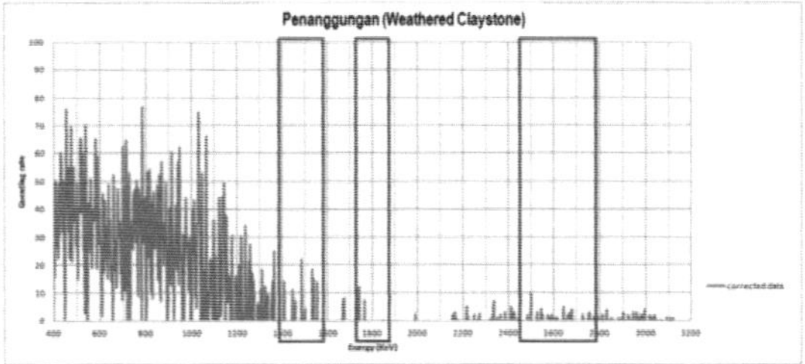

Potassium Uranium Thorium

(d)

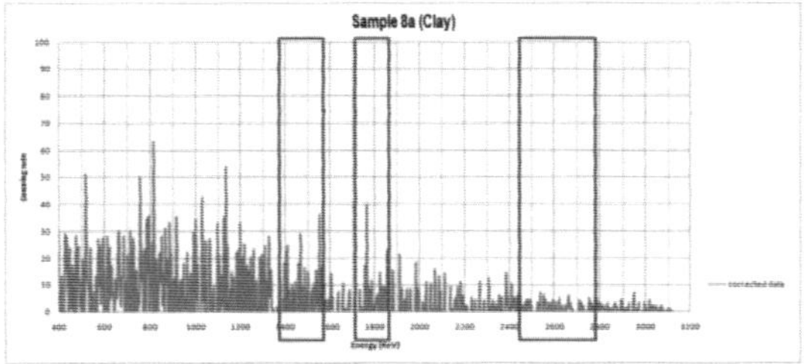

Potassium Uranium Thorium

(e)

IOP Conf. Series: Journal of Physics: Conf. Series **799** (2017) 012027 doi:10.1088/1742-6596/799/1/012027

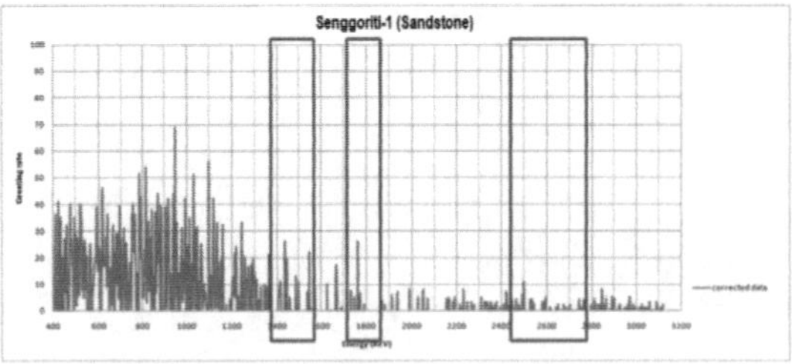

Potassium Uranium Thorium

(f)

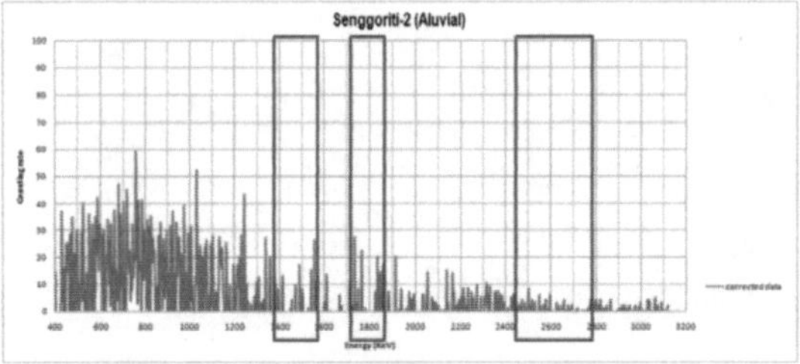

Potassium Uranium Thorium

(g)

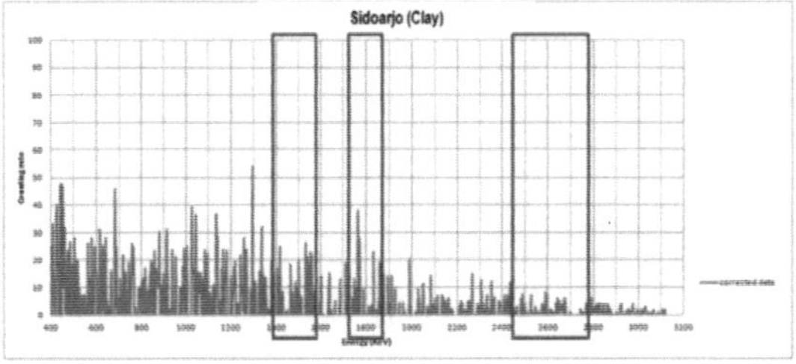

Potassium Uranium Thorium

(h)

Fig. 3. Counting curve from (a) Weathered Claystone Cangar-3 (b) Claystone Jolotundo (c) Tuff Welirang (d) Weathered Claystone (e) Clay Sample 8A (f) Sandstone Senggoriti-1 (g) Alluvial Senggoriti-2 (h) Clay Sidoarjo.

Table 1. Total Countingrate of Natural Gamma Radiation

Sample	Countingrate		
	Potassium	Uranium	Thorium
Welirang Tuff	596	117	360
Jolotundo Clay	496	122	306
Cangar-1 Claystone	413	125	191
Sample 8a Clay	345	123	215
Sidoarjo Clay	336	125	222
Senggoriti-1 Sandstone	192	64	56
Senggoriti-2 Aluvial	170	80	154
Cangar-2 Sand	140	87	87
Penanggungan Weathered Claystone	161	58	19
Cangar-3 Weathered Claystone	79	66	13

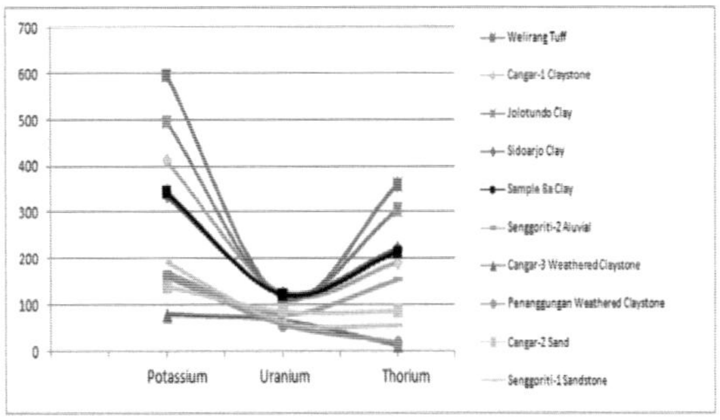

Fig. 4. Total Coutingrate Vs Natural Gamma ray (Thorium, Uranium, Potassium)

4. Conclusions

The difference of gamma ray spectra among Sandstone, Claystone, Tuff, Alluvial, Weathered clay, Shale/clay is significant. The rock sample collected from East Java Basin has lower Uranium and Thorium's spectra compared than Postassium's spectra. Weathered vulcanic clay has lower counting rate of gamma ray than alluvial deltaic clay counting rate. Therefore, gamma spectrometrometry can be used as tool for characterizing the enviroment of clay whether vulcanic or alluvial-deltaic. This phenomena indicates that gamma ray spectrometry can be as tool for characterizing the clay whether it tends to Smectite or Illite.

Acknowledgments

This research is fully supported by Indonesian ministry of education research grant and Rock Fluid Imaging Lab.

References

[1]. M. Tzortzis, H. Tsertos, S. Christofides, G. Christodouildes, *Gamma radiation measurements and dose rates in commercially-used natural tiling rocks (granites)*, J. of. Enviromental Rad, **70**, issue 3, 2003, pp 223-235.

[2]. L. LQvborg, P. Nyegaard, E. M. Christiansen, and B. L. Nielsenz, 1. *Borehole logging for uranium by gamma-ray spectrometry,* Geophysics, **15**, no. 6, 1980, pp 1077- 1090.

[3]. R. B. K. Shives, B. W. Charbonneau, K. L. Ford, *The detection of potassic alteration by gamma-ray spectrometry—Recognition of alteration related to mineralization*, GEOPHYSICS, 65, no. 6, 2000, pp. 2001-2011

[4]. M. Tzortzis, H. Tsertos, S. Christofides, G. Christodouildes, *Gamma-ray measurements of naturally occurring radioactive samples from Cyprus characteristic geological rocks*, Rad. Measurements, **37**, 2003, pp 221-229

[5]. M. Zubair, *Measurement of natural radioactivity in rock samples using gamma ray spectrometry*, Rad. Prot. and Environment, **38**, 2015, 11- 13.

[6]. K. Jeevarunka, G. Sankaran, P. S. Hameed, R. Mathiyarasu, *Evaluation of natural gamma radiation and absorbed gamma dose in soil and rocks of Perambalur district (Tamil Nadu, India)*, J. of Rad. and Nucl. Chem, **302**, Issue 1, (2014), pp 245-252

[7]. J. A. S. Adam, G. E. Fryer, *Portable gamma Ray Spectrometer for field determination of Thorium, Uranium, and Potassium,* Natural Radiation Environment, pp 577-96, University of Chicago Press 1964.

[8]. H. W. Taylor, *Gamma rays emitted by uranium and thorium in the energy range 10–120 keV*, Int. J. of Appl. Rad. and Isotopes, **24**, issue 10, (1973) pp 593-597

[9]. Barnes P. *Structure and performance of cements*. Applied Science Publishers. London. 1983.

[10]. Peterson RS. *Experimentalgamma ray spectroscopy and investigation of environmental radioactivity*. Spectrum Techniques. 1996.

IOP Conf. Series: Journal of Physics: Conf. Series **799** (2017) 012028 doi:10.1088/1742-6596/799/1/012028

Subsurface Temperature Modeling using Integrated Modeling for Nuclear Reactor Site Assessment in Volcanic Zone

Bagus Endar B. Nurhandoko[1, 2], Rizal Kurniadi[1], Elfa Fatiah[2], Muhammad Rizal Abda[2], Rio Martha[2] , Sri Widowati [3]

[1]*Physics Department, Faculty of Mathematics and Natural Science,Institut Teknologi Bandung*
[2]*Rock Fluid Imaging Lab.*
[3]*School of computing, Telkom University*

Email: bagusnur@bdg.centrin.net.id

Abstract Indonesia has giant vulcanic arc that almost the largest vulcanic arc in the world. Therefore, one of the main risk for nuclear site plan is the vulcanic area. Therefore to reduce the risk, one of most safety nuclear site plant is old vulcanic area.In this paper, we propose to predict subsurface temperature profile to ensure the condition of subsurface of vulcanic zone. Geothermal heat flow is important parameter in modeling of subsurface temperature. The subsurface temperature is one of vulcanic activity parameter which very important for nuclear site plant risk assesment.The integrated modeling for predicting subsurface temperature profile is carried out by combining geothermal heat flow and subsurface profiles resulted from either seismic or gravity measurement. The finite difference of Fourier's law is applied to surface temperature, temperature gradient, geothermal heat flow and thermal conductivity profile for producing subsurface temperature distribution accurately. This subsurface temperature profile is essential to characterize the vulcanic zone whether it is still active or inactive. Characterization of vulcanic activity is very useful to ensure or to minimize the risk of nuclear site plant in vulcanic zone. One of interesting case study of nuclear site plan in Indonesia is mount Muriah site plan, this method is useful to ensure whether mount Muriah is still active or inactive now

1. Introduction

The site plan assessment is one of important study in nuclear plant development. The nuclear site plan assesment in Vulcanic area covers history of seismic activity [1, 2, 3], volcanic hazards [4], and geological evaluation [5].

Indonesia has potential hazards of volcanism and seismicity since lies on the world's major chains of active volcanoes and has densely populated region.

Seismic hazard assessment needs the information of regional tectonics and fault characteristics properly. The successfull assessment of volcanic hazard strongly depends on quality of sub surface research such as geologic and geothermal researches. Some authors have been investigated about Muria volcano [6, 7, 8, 9].

One of important study in volcanic hazard is a prediction whether any this volcano might be capable of future activity affecting the nuclear plant. The concept of an activity risk of volcano likes a activity risk of fault. Activity risk can be used to distinguish those features whether has potential hazardous or not.

In the case of Muria Genuk complex Indonesia, the initial studies took two forms [10]. Firstly, the record of volcanism was examined by determine whether any of the volcanic centers had historical eruptive activity. The search revealed no record of eruptions from any volcano on Muria Peninsula. Secondly, the volcanoes and an area of 100 km^2 around the site were examined to determine whether there were any manifestations of current magmatic activity. Seismic records were examined for evidence of greater than normal seismicity. Temperature gradients of 5-6 ^{o}C per 100 m were measured in boreholes extending to depths of about 250 m near the site. These temperature gradients were considered normal for the region.

In this work, we propose a technique to obtain temperature distribution of subsurface using surface temperature measurement and temperature gradient.

2. Experimental Method

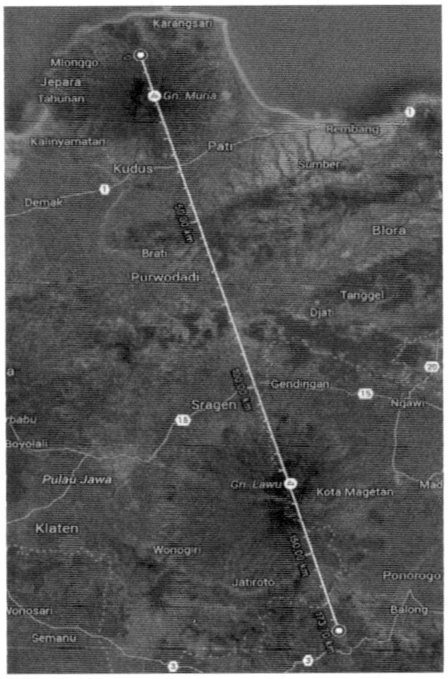

Figure 1. The Line of temperature measurement

Figure 1 shows the part of East Java map that indicate the position of temperature modeling. Spacing of modeling about 10 km. Beside temperature, thermal conductivities are measured.

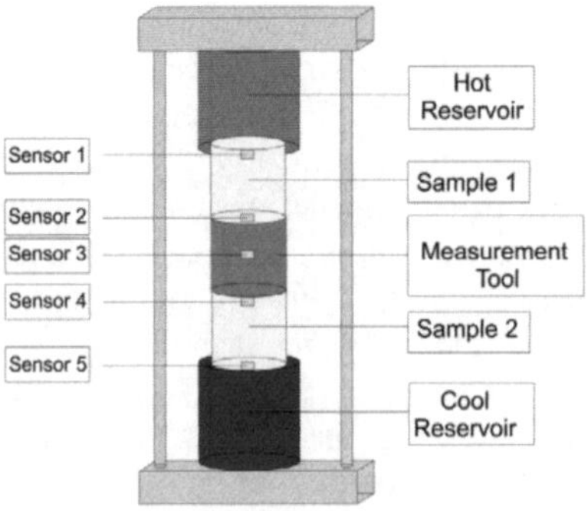

Figure 2. ThermConduct™

The modeling uses data of temperature measured at line in East Java Basin, from northern part to southern part of East Java, similar direction with the line shown in Figure 1, the true position is around 150 km to the east from line shown in Figure 1. The temperature data and subsurface geologi model beneath line shown in Figure 1 are input of subsurface temperature modeling.

The temperature data collected from water temperature from various depth of water level. The conductivity of samples are measured by ThermConduct™ which equipped by 5 sensors that they attach at sample (rock and cooper) and 2 thermal reservoirs (top and bottom) as shown in Figure 2.

Sample preparation includes coring, smoothing, dimension measurement and density measurement. Thermal conductivity measurement needs 2 samples, sample has around 3-4 centimeter in length and 3 cm in diameter.

Temperature sensors are attached at samples as shown by figure 2. The water in hot reservoir is boiled and kept temperature in constant value. The data (time and temperature) are collected via temperature sensor in real time.

The conductivity measurement uses Fourier equation as following:

$$\frac{Q}{A} = K\frac{dT}{dz} \quad (1)$$

$$Q = m.c.\Delta t \quad (2)$$

Q = Total caloric that pass trough
A = Cross sectional area of sample
K = Rock conductivity
$\frac{dT}{dz}$ = Gradient temperature of sample.
m= mass of copper
c= Thermal capacity of cooper
Δt=temperature differentiation of copper

The modeling uses various sample of lithologies: Carbonate (fig. 4), Clay (fig. 5), Tuff (fig. 6), Andesite (fig.7), and also sand sample.

Figure 8, 9, 10, 11, 12, 13, 14, 15 show measurement result of conductivity, we can estimate simply the value of conductivity from average points of measurement.

Subsurface temperature modeling use conductivity information which confirm with geological model and surface temperature data, flowchart of modeling is shown by figure 3.

The subsurface temperature profile can be estimated by combining surface temperature, temperature gradient, thermal conductivity, and geological model that it was related with gravity, seismic, and well data.

The forward model of temperature distribution is carried out thru Fourier Law which solved numerically by finite difference.

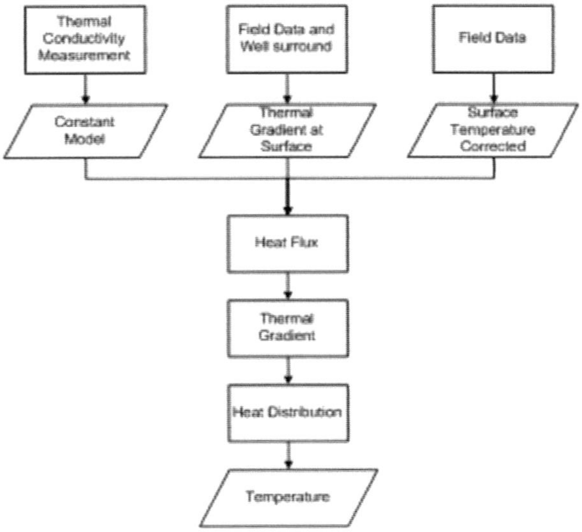

Figure 3. Flow chart data peocessing of subsurface temperature distribution

3. Results and Dicussion

Subsurface geology model is carried out by gravity modeling constraint by geology map and well information.

Figure 4 shows lithology model resulted from gravity modeling which adjusted by considering information from geological map and wells. Colors in figure 4 indicate dominant lithology. This information of lithology is tabulated in table 1.

Intrusion rock is shown in northern part where position of Mt. Muriah, in the middle part the thrust fault creates uplitft of basement. Mt. Lawu shows anomaly due of Bouguer anomaly is negative which indicates relatively low density. This anomaly may be correlated by lithology of crater whether dominatedby sand and boulder or fractured lithology.

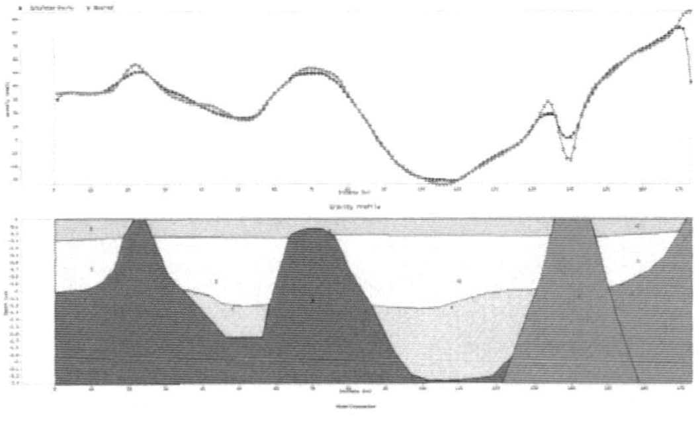

Figure 4. Subsurface geology model beneath line of figure 1 by gravity modeling.

Table 1. Density of rocks formation

Color	Formation	Delta Density	Density
Red	Quarter Vulcanic	1.07	3.4
Purple	Basement	0.54	2.87
orange	Volcanic Lawu	-0.22	2.11
Purple	Eosen-Oligosen sediment	0.27	2.6
Yellow	Miosen sediment	0.07	2.4
Green	Plio-Plistosen Sediment	-0.46	1.87

Table 2. The tabulation of "dummy" surface data

Daerah	Position (Km)	Temp °C	Gradient (C/ 100 m)	Temp Correction °C
Mlonggo	0	30.48	5.5	31.52
Brati	55	30.74	3.6	31.06
Purwoda di	70	29.64	3.7	29.97
Gending an	100	29.9	3.78	30.05
Sragen	105	29.92	2.7	30.51
Magetan	135	29.79	2.05	29.95
Jatiroto	155	29.1	2.5	29.45

Figure 4. Carbonate sample

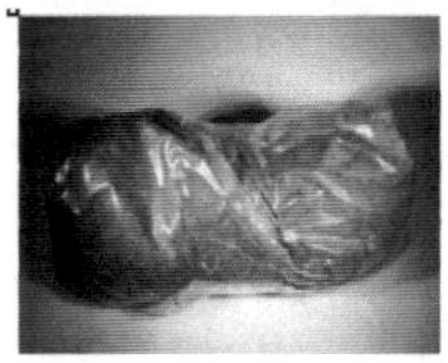

IOP Conf. Series: Journal of Physics: Conf. Series **799** (2017) 012028 doi:10.1088/1742-6596/799/1/012028

Figure 5. Clay sampel

Figure 6. Tuff Sampel

Figure 7. Andesit

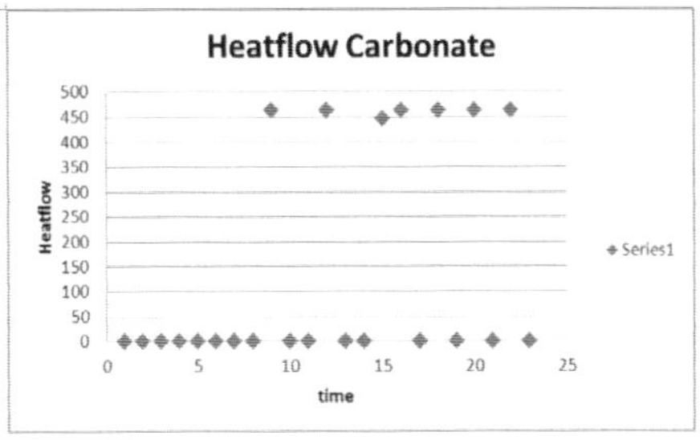

Figure 8. Carbonate Heatflow

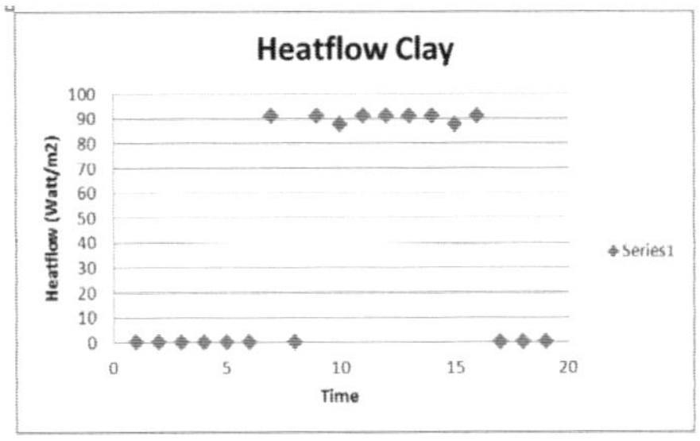

Figure 9. Clay Heatflow

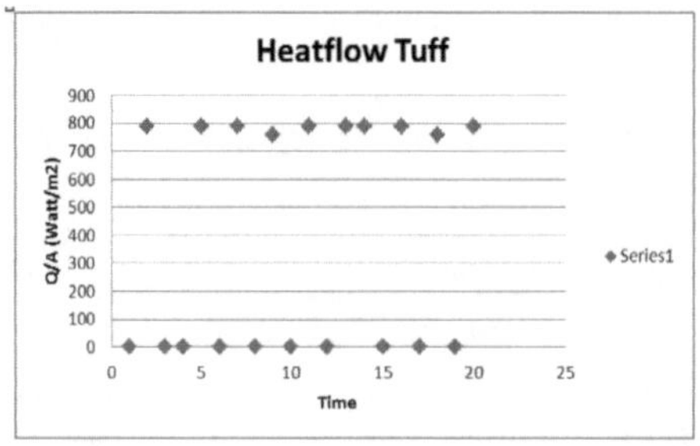

Figure10. Tuff Heatflow

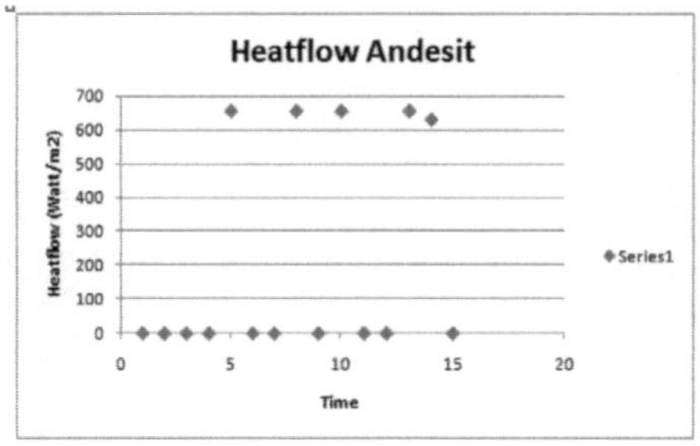

Figure 11. Andesit Heatflow

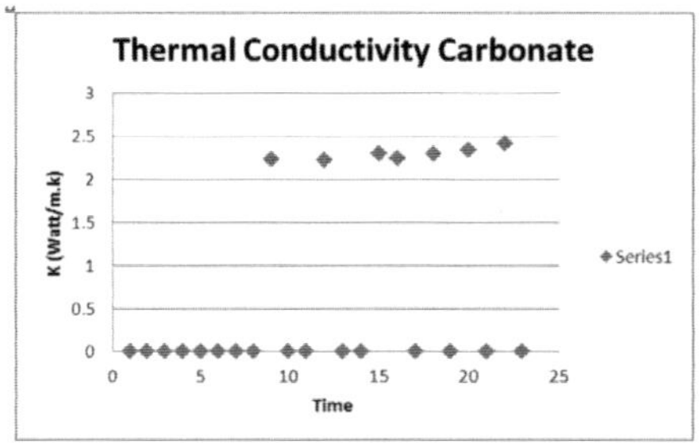

Figure 12. Carbonate thermal conductivity

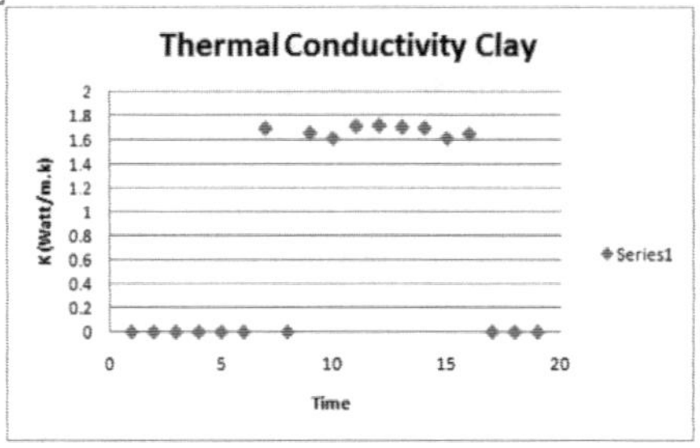

Figure13. Clay thermal conductivity

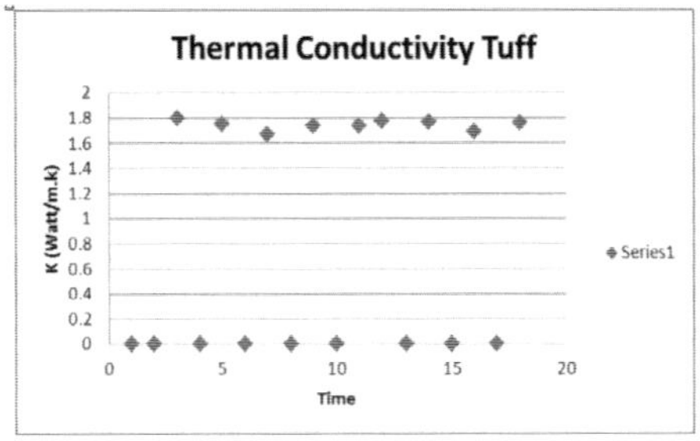

Figure 14. Tuff thermal conductivity

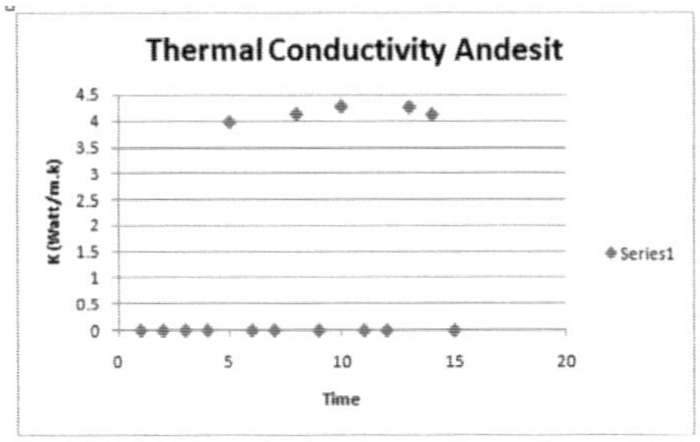

Figure15. Andesit thermal conductivity

Thermal conductivity (K) of material is unique property. From results, K value is increasing due to increasing of mass density. Hi density rock has small pores hence there are small volume of fluid in the rock. Because of small amount of fluid, the heat is easier to be transferred.

Figure 16 shows subsurface conductivity model calculated based on geology model from gravity modeling as shown by Figure 4.

By combining conductivity model of figure 16 and surface temperature data including geothermal heat flow, the subsurface temperature profile can be resulted as shown by Figure 17.

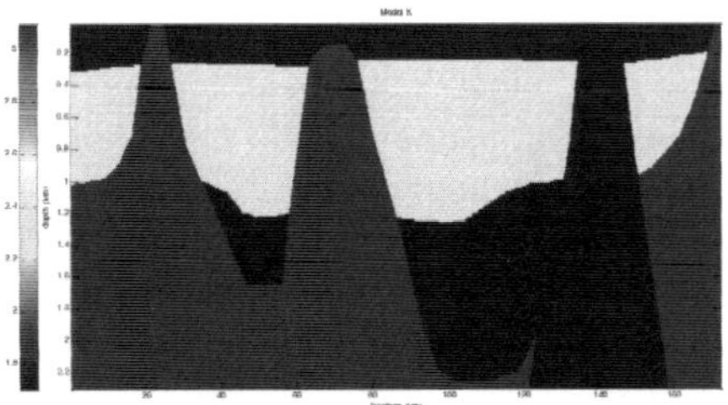

Figure 16. Subsurface thermal conductivity model

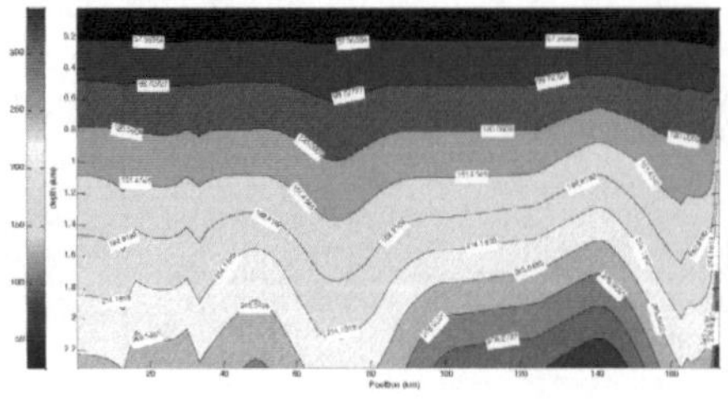

Figure 17 Sub surface temperature beneath line as shown by figure 1.

4. Conclusions

1. The input data for subsurface temperature profiling are surface temperature, temperature gradient, conductivity of rock and subsurface geology modeling. The subsurface geology model is calculated from gravity modeling which constraint by surface geology, well data and seismic.
2. The proposed heatflow measurements as well as subsurface temperature profiling can provide robust estimation of subsurface temperature profile.
3. The subsurface temperature profile represents condition of sub-volcanic zone for determining the activity of volcanic zone. It is very important for nuclear site plan assessment.

Acknowledgments
This research is fully supported by Indonesian ministry of education research grant and Rock Fluid Imaging Lab.

References

[1]. R. P. Kennedy and M. K. Ravindra, *Seismic fragilities for nuclear power plant risk studies*, Nucl. Egineering and Design, **79**, issue 1, 1984, pp 47-68.
[2]. S. Desai, D. Choudhury, *Site-Specific Seismic Ground Response Study for Nuclear Power Plants and Ports in Mumbai*, Nat. Hazards Rev., 10.1061/(ASCE)NH.1527-6996.0000177, 04015002.
[3]. K. Decker, L. Guerrieri, Y. Fukushima, *The Contribution of Paleoseismology to Seismic Hazard Assessment in Site Evaluation for Nuclear Installations*, EGU General Assembly 2015, held 12-17 April, 2015 in Vienna, Austria. id.11346.
[4]. A. M. F. Lagmay, et.al, *Geological hazards os SF Natib Volcano, site of the Bataan Nuclear Power Plant, the Philippines*, Geological Society, London, Special Publication 2012, p 151-169.
[5]. M. J. Adair, *Geologic evaluation of a site for a nuclear power plant*, Rev. in Eng. Geology, **4**, 1979, p 27-40
[6]. van Bemmelen, R.W., *The Muriah Volcano (Central Java) and the origin of its leucite-bearing rocks*. K. Ned. Akad. Wet. **50**, 1947, pp 653-658
[7]. Boomgaart, L., *Some data on the Muriah volcano (Java) and its leucite-bearing rocks*. K. Ned. Akad. Wet. **50**, 1947, pp 649-652
[8]. Maury, R.C., Soeria-Atmadja, R., Bellon, H., 1987. *Nouvelles donne¤es ge¤ologiques sur les deux association magmatiques du volcan Muria (Java, Indonesia)*. C. R. Ser. 2 **304**, 1987, pp 175-180
[9]. Nicholls, I.A., Whitford, D.J., 1983. *Potassium-rich volcanic rocks of the Muriah complex, Java, Indonesia; products of multiple magma sources?* J. Volcanol. Geotherm. Res. **18**, 1983, pp 337-359
[10]. A.R. McBirney et al. *Volcanic and seismic hazards at a proposed nuclear power site in central Java*, Journal of Volcanology and Geothermal Research **126** (2003) pp 11-30

Preliminary Study of Gas Cooled Fast Breeder Reactor with Heterogen Percentage of Uranium–Plutonium Carbide based fuel and 300 MWt Power

Sandro Clief Pattipawaej[1], **Zaki Su'ud**
Nuclear Physics and Biophysics Research Division, Physics Department, Faculty of Mathematics and Natural Science, Bandung Institute of Technology Indonesia, Jalan Ganesha 10 Bandung 40132, INDONESIA

E-mail: sandro_clief@yahoo.com

Abstract. A preliminary design study of GFR with helium gas-cooled has been performed. In this study used natural uranium and plutonium results LWR waste as fuel. Fuel with a small percentage of plutonium are arranged on the inside of the core area, and the fuel with a greater percentage set on the outside of the core area. The configuration of such fuel is deliberately set to increase breeding in this part of the central core and reduce the leakage of neutrons on the outer side of the core, in order to get long-lived reactor with a small reactivity. Configuration of fuel as it is also useful to generate a peak power reactors with relatively low in both the direction of axial or radial. Optimization has been done to fuel fraction 45.0% was found that the reactor may be operating in more than 10 year time with excess reactivity less than 1%.

1. Introduction

One of the problem in nuclear reactor is fuel waste. This fuel waste was remainder of LWR (Light Water Reactor) which usually keep in the save place or being processed to become the new fuel. Waste of LWR has been processed can be used as fuel in fast reactor (FBR). This is one of the advantage of fast reactor compared to thermal reactor.

In addition it can be used to burn LWR waste, FBR gas cooled (Gas Cooled Fast Reactor) also has another advantage that allows it to produce the output temperature is more than 800^0C with a relatively long operating life and small excess reactivity [1].

Disadvantages of this type of reactor is safety factor. High conversion ratio value and large breeding rate of reactor impacts the reactor stability. Some neutronics parameters be related with reactor stability is reactivity and power distribution.

2. Design Concept and Calculation Methods

In this study, the reactor core is designed three-dimensional hexagonal compiled by hexagonal shaped assemblies with exactly the same size. Core reactor consists of three important parts of active cores located in the centre of core, then covered with reflectors and a blanket stand beside the outer core [2]. Active core in the radial direction is divided into three regions. Region I consist of 3 line assembly, region II and III consist of 2 line assembly. Region I is filled by twelve fuel assembly (A), one inner

---

[1] To whom any correspondence should be addressed.

shut down assembly (D) and six inner control rod assembly (F). Region II is filled by thirty six intermediate fuel assembly (B) and six outer shut down (E). Region III is filled by forty eight fuel assembly (C) and eighteen outer control assembly (G).

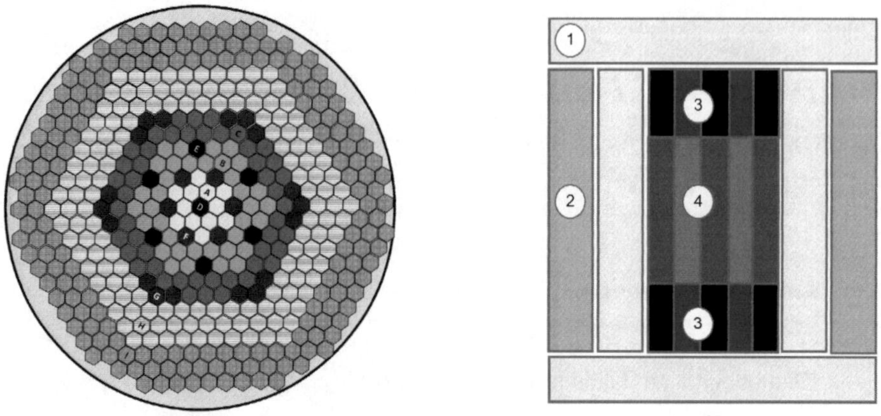

(a) (b)

Figure 1. Core design (a) radial direction, (b) axial direction

In the axial direction of active core is divided into three regions (Top, middle and bottom). Central region has a size larger than the upper or bottom region which has the same size. A general description of the reactor core in radial and axial directions given in 'figure 1 (a) and (b)' [2].
Each fuel assembly in active core is filled by 127 cylindrical fuel pin [2], while the control and shutdown is filled by 108 cylindrical fuel pin plus one control or shut down rod with a much larger size.

Table 1. General specification of reactor

Parameter	Specification
Thermal Power	300 MWth
Refueling period	10 – 20 years
Core Geometry	Hexagonal 3D
Coolant	Helium gas
Number of fuel assembly:	
(Region I, II, III)	(12, 36, 48)
Number of control Rod assembly	24
Number of shut down assembly	7
Pellet material	(U-Pu)C
Pellet diameter (mm)	7.5
Cladding thickness (mm)	0.94
Cladding material	SiC
Uranium enrichment	^{235}U(0.72%),^{238}U(99.28%)
Plutonium enrichment	^{238}Pu(1.7%), ^{239}Pu(58.8%), ^{240}Pu(22.8%), ^{241}Pu(1.7%), ^{242}Pu(5%) [3]

The fuel used in core reactor is uranium plutonium carbide. In core, the fuel each region division based on the high percentage of plutonium (Pu). The percentage of Pu in region III (radial direction) and region top/bottom (axial direction) is made higher to reduce leaking neutron. Low percentage of

Pu in region I (radial direction) and axial center region contributes to the high value of conversion ratio. General specification of reactors are given in 'table 1'.

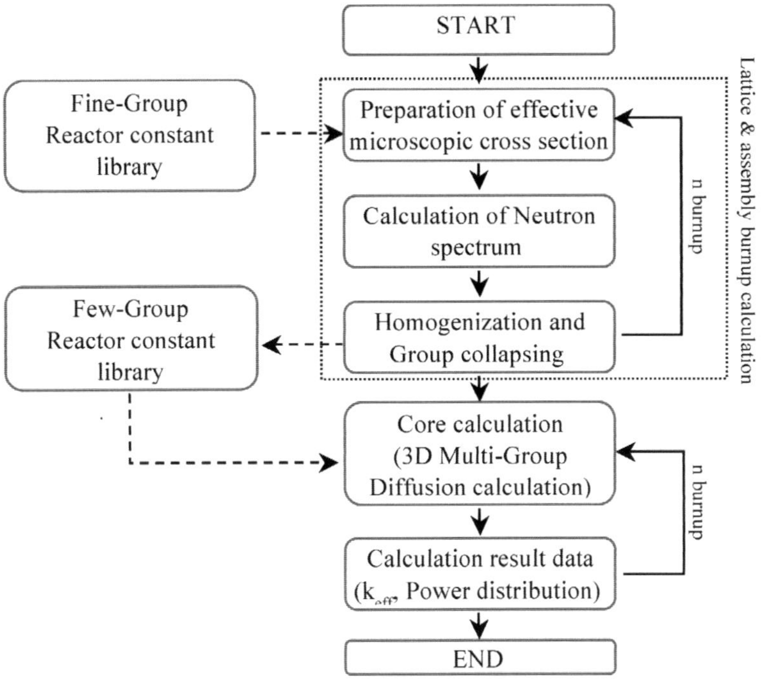

Figure 2. Block diagram of neutronic calculation with SRAC [4]

Neutronic analysis of reactor core analyzed used SRAC (Standard Reactor Analysis Code) code system. In SRAC calculation is divided into two parts, namely lattice calculation and core calculation. Macroscopic cross section is calculated at lattice calculation, based on some energy group utilizing reactor constant library JENDL-32 [5]. Furthermore macroscopic data, can be used in 3D diffusion equation. Solution of this equation generated in the core calculation. Diagram block of design calculation is given in 'figure 2'.

3. Results and Discussion

Optimization is done for fuel fraction 45.0%. For this fuel fraction obtained maximum configuration of Pu in each region is given by Reg I: 9.2%, Reg II: 11.4%, Reg III: 15.9% for up/bottom region and Reg I: 7.2%, Reg II: 9.4%, Reg III: 13.9% for middle region in axial direction.

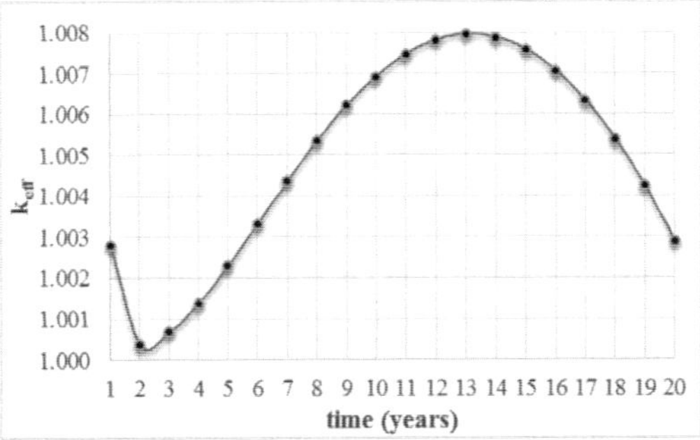

Figure 3. Effective multiplication factor (k_{eff})

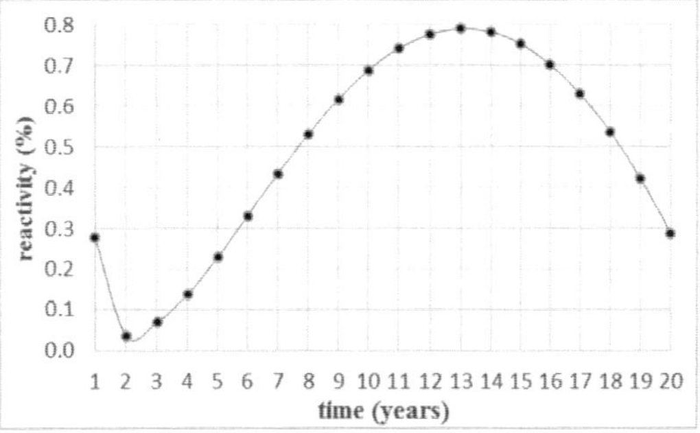

Figure 4. Reactivity

The value of the the effective multiplication factor (k_{eff}) and excess reactivity of reactor is given by 'figure 3' and 'figure 4'. While the power distribution in radial and axial direction is given by 'figure 5'.
From 'figure 3 and figure 4', it appears that the reactor has operation time to 20 years with reactivity less than 0.8%. In the beginning of life (BOL) at burn up step 1 and 2, the reactor is dominated by burning process. From step 3 until step 13 is dominated by breeding process. In end of life (EOL), reactor had a positive reactivity that allows the reactor to operate more than 20 years.

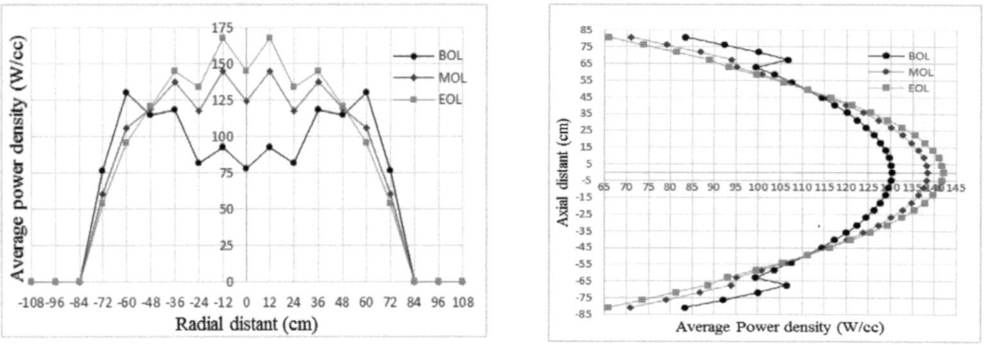

(a) (b)

Figure 5. Power distribution of reactor (a) radial direction, (b) axial direction

The large time of reactor operation is caused by the amount of fuel in this reactor. The other reason is the reactor have a large value conversion ratio and has low percentage of Pu (high fertile material number) at each fuel region. Conversion is high and the number of fertile material which much guarantee the process of breeding for a long time.
From 'figure 5', it appears that the reactor had minimum power at control and shut down assemblies due to the amount of fuel material is less. The minimum power is also found in Region III (radial direction) and top/bottom (axial direction) due to the number of neutrons leaking more than any other region.
The transition of power distribution in the reactor was also influenced by the high value conversion ratio. The rate of transition conversion ratio or in this case a high breeding rate especially in area with the number of neutrons and fertile material is abundant, have a direct impact on the rate of transmission in the region. Base on "figure 5", shows that the power transition is a significant change from the first year (BOL) to tenth years (MOL). Furthermore, after MOL power transition and breeding rate is also declining.

4. Conclusion

Placement of fuel material with a low percentage of plutonium in the centre of the reactor and the high neutron population in this area impact on high internal convertion of reactor. High internal convertion and high fuel fraction (45.0%) guarantees a long life of the reactor. This reactor can operate more than 20 years with reactivity less than 0.8% but high transition of power distribution. The rate of transition power distribution from BOL to MOL is larger than from MOL to EOL caused by from BOL to MOL reactor is dominated by breeding process and from MOL to EOL is dominated by burning. The heterogeneous of fuel material in any directions able to cut peak power of reactor. Power peaking factor of reactor form BOL to EOL is less than 1.3.

References

[1] Elder R and Allen R 2009 Nuclear heat for hydrogen production: Coupling a very high/high temperature reactor to a hydrogen production production plant *Progress In Nuclear Enegy* **51** pp 500-525

[2] Waltar A E, Todd D R and Tsvetkof P V 2012 *Fast Spectrum Reactors* (London: Springer) p 487, 28

[3] Expert group 1999 Actinide and Fission Product Partitioning and Transmutation, Nuclear Energy Agency (NEA)/ (OECD), -, p 303

[4] Oka Y, Madarame H and Uesaka M 2014 *An Advance Course in Nuclear Engineering* (Tokyo: Springer) p 61

[5] Okumura K 2002 *SRAC version 2002* (Japan Atomic Energy Research Institute) pp 1-28